Water Science and Technology Library

Volume 90

The aim of the *Water Science and Technology Library* is to provide a forum for dissemination of the state-of-the-art of topics of current interest in the area of water science and technology. This is accomplished through publication of reference books and monographs, authored or edited. Occasionally also proceedings volumes are accepted for publication in the series. *Water Science and Technology Library* encompasses a wide range of topics dealing with science as well as socio-economic aspects of water, environment, and ecology. Both the water quantity and quality issues are relevant and are embraced by *Water Science and Technology Library*. The emphasis may be on either the scientific content, or techniques of solution, or both. There is increasing emphasis these days on processes and *Water Science and Technology Library* is committed to promoting this emphasis by publishing books emphasizing scientific discussions of physical, chemical, and/or biological aspects of water resources. Likewise, current or emerging solution techniques receive high priority. Interdisciplinary coverage is encouraged. Case studies contributing to our knowledge of water science and technology are also embraced by the series. Innovative ideas and novel techniques are of particular interest.

Comments or suggestions for future volumes are welcomed.

Vijay P. Singh, Department of Biological and Agricultural Engineering & Zachry Department of Civil Engineering, Texas A and M University, USA
Email: vsingh@tamu.edu

More information about this series at http://www.springer.com/series/6689

Agnieszka Stec

Sustainable Water Management in Buildings

Case Studies From Europe

 Springer

Agnieszka Stec
The Faculty of Civil and Environmental
Engineering and Architecture, Department
of Infrastructure and Water Management
Rzeszow University of Technology
Rzeszów, Poland

ISSN 0921-092X   ISSN 1872-4663 (electronic)
Water Science and Technology Library
ISBN 978-3-030-35961-4   ISBN 978-3-030-35959-1 (eBook)
https://doi.org/10.1007/978-3-030-35959-1

This Springer imprint is published by the registered company Springer Nature Switzerland AG
The registered company address is: Gewerbestrasse 11, 6330 Cham, Switzerland

Preface

Water is a key resource of nature and it is indispensable for human health and life. Water resources determine the proper functioning of sectors such as agriculture, industry, tourism, transport, and energy. Many regions in the world, including Europe, are affected by the problem of limited water availability. Freshwater resources are particularly important as their quality and quantity are constantly decreasing. Climate changes, urbanization, and population growth are expected to increase water shortages in the future causing many serious adverse environmental changes threatening current and future generations. Therefore, it is important that the management of this resource is sustainable, which requires an integrated approach taking into account the environmental, economic, and social dimensions. The European Union Water Framework Directive rightly reminds us that "*water is not a commercial product like any other but, rather, a heritage which must be protected, defended and treated as such.*"

Considering the above, the monograph presents some possibilities of saving fresh water by using alternative sources of water in residential buildings. The attention was drawn to rainwater and gray water, which are increasingly used around the world and are a valuable source of water, especially for non-potable uses. The main purpose of the book was to show that unconventional water systems not only reduce the consumption of tap water, but can also be financially profitable. Therefore, some analyses of Life Cycle Cost rainwater harvesting systems and graywater recycling systems located in selected European cities were carried out. These locations were characterized by different climatic conditions, different water consumption by residents, and different water purchase prices. This allowed determining the impact of these parameters on the profitability of using the tested systems.

Alternative sources of water in residential buildings, in spite of the development and implementation of appropriate ecological strategies, are used quite rarely in many European countries. This can be influenced by many factors, including the financial aspect, hygiene, and legal considerations. To find out the reason for this, a survey was conducted in selected European countries and its results are presented in this monograph. Such research is very important since the acceptance of society can be decisive in implementing unconventional solutions in construction.

The research outcomes presented in the monograph are not only scientific but also practical ones and can be a guide for potential investors in the decision-making process already at the investment planning stage. This book, therefore, can contribute to the growing interest in alternative water systems in housing, and thus to promote sustainable construction.

Rzeszów, Poland   Agnieszka Stec

Acknowledgements

The author would like to thank the reviewers, namely Prof. Ing. Zuzana Vranayová, Ph.D., Professor at the Department of Building Facilities at Faculty of Civil Engineering, Technical University in Košice and Prof. Ing. Štefan Stanko, Ph.D., Professor at the Department of Sanitary and Environmental Engineering at Faculty of Civil Engineering, Slovak University of Technology in Bratislava.

The author would like to thank also the publisher—Springer Nature for providing the opportunity for this publication.

Contents

Abbreviations

BOD$_5$	Biochemical oxygen demand
EEA	European Environment Agency
EU	European Union
EUR	Euro
EurEau	European Federation of National Associations of Water Services
GWHS	Graywater recycling system
IRR	Internal Rate of Return
LCC	Life Cycle Cost
MBR	Membrane bioreactor
NPV	Net Present Value
PB	Payback Period
RBC	Rotating biological reactor
RWHS	Rainwater harvesting system
SBR	Sequencing batch reactor
WFD	Water Framework Directive
WHO	World Health Organization
WWAP	World Water Assessment Programme
YAS	Yield-after-spillage algorithm
YBS	Yield-before-spillage algorithm

Chapter 1
Introduction

Currently, the world is facing serious environmental problems resulting from climate changes, population growth, urbanization, and global warming. In addition, the constantly growing demand for various types of raw materials causes excessive exploitation of natural resources. Therefore, sustainable exploitation and conservation of these resources are crucial for the contemporary development and existence of future generations (López-Morales and Rodríguez-Tapia 2019; Urbaniec et al. 2017). To achieve this, it is necessary to implement pro-ecological strategies based on alternative technologies in all areas of the economy, including housing, which has a significant demand for water and energy. It is estimated that domestic use of fresh water is about 10% of the total global water demand (Bocanegra-Martínez et al. 2014).

1.1 Background

Human activities, mainly urbanization and industrialization, contribute to an increase in the degree of environmental pollution causing, among others, deterioration of the quality of water resources and limitation of their availability (Salvadore et al. 2015; Vörösmarty et al. 2000). Currently, almost 54% of the world's population live in urban areas, and this percentage is expected to increase to 66% in 2050 (UN 2014). The increase in population migration to cities necessitates their expansion, causing, among others, disorders in the natural hydrological cycle (Oudin et al. 2018). Not without significance for the state of water resources is also excessive and unconscious water consumption, which is also observed in residential buildings, where over 50% of water is used for purposes where the quality of potable water is not required. Meeting the needs of an ever-growing population increases pressure on natural resources, including water resources, leads to water deficits in many regions (Mekonnen and Hoekstra 2016; Ercin and Hoekstra 2014; Vörösmarty et al. 2000). Given the current

© Springer Nature Switzerland AG 2020
A. Stec, *Sustainable Water Management in Buildings*,
Water Science and Technology Library 90,
https://doi.org/10.1007/978-3-030-35959-1_1

annual growth of the world population at 1.1% resulting in an increase of approximately 83 million people per year, the human population is projected to reach 9.8 billion in 2050 (UN 2017). According to some forecasts, this will increase the global water demand by 55% this year (WWAP 2015).

In searching for alternative water sources, special attention was paid to rainwater, which is characterized by relatively low pollution, especially those flowing from the roofs of buildings, which does not require advanced treatment processes (Şahin and Manioğlu 2018; Zhang and Hu 2014). Rainwater harvesting systems (RWHS) have been used for many years around the world and are perceived as one of the strategies enabling adaptation of the water management sector to the changing climate and as a response to the constantly increasing demand for water (Zhang et al. 2019; Mwenge Kahinda et al. 2010; Pandey et al. 2003). Rainwater as an alternative source of water in buildings is used both as potable and non-potable water (Fewkes 2006; Campisano et al. 2017). The task of RWHS is to collect, pre-clean, collect, and then use water for various purposes in buildings (Silva et al. 2015; Jha et al. 2014). In regions where water scarcity is not so noticeable, rainwater harvesting systems are used for non-potable use, most often in hybrid systems as systems supplementing a traditional water source. In these cases, treated rainwater is primarily used for toilet flushing (Jones and Hunt 2010; Słyś et al. 2012; Kaposztasova et al. 2014), irrigation of green areas (Devkota et al. 2015), cleaning works, washing (Morales-Pinzón et al. 2014; Imteaz et al. 2012), car washing (Ghisi et al. 2009), and irrigation of crops (Unami et al. 2015). In developing countries and areas where centralized water supply systems are not feasible, rainwater is often used as potable water (Lee et al. 2017).

The interest in RWHS is also increasing in regions, where there are no significant problems with access to tap water, and this increase is caused by a social trend promoting a sustainable lifestyle that is supposed to limit people's ecological footprint (Hofman-Caris et al. 2019).

The possibilities of using rainwater in various types of buildings were widely explored worldwide. For instance, in single-family homes (Severis et al. 2019; Martin et al. 2015), multistory residential buildings (Ghisi and Ferreira 2007; Słyś and Stec 2014), office buildings (Yana et al. 2018; Ward et al. 2012), schools (Lee et al. 2017; Cheng and Hong 2004), dormitories (Stec and Zeleňáková 2019), sports facilities (Zaizen et al. 1999), hospitals (Lade and Oloke 2017), airports (Moreira Neto et al. 2012), and gas stations (Ghisi et al. 2009).

Depending on the location and the type of the building, climatic conditions, the size of the drained area, and the demand for water, saving of tap water can reach different levels thanks to the application of RWHS (Amos et al. 2018; Bashar et al. 2018; Imteaz et al. 2011, 2014). Zhang et al. performed simulation tests of the use of rainwater and wastewater in residential buildings of the city of Cranbrook in Western Australia. The tap water saving was 32.5% when reusing gray water and 25.1% when using rainwater for toilet flushing and garden irrigation (Zhang et al. 2010). Similar analyses were performed by Coombes et al. (1999) for another Australian city. In the study, they took into account 27 apartment buildings located in Newcastle. It turned out that for the data adopted for the analysis, rainwater could cover 60% of the water demand in these buildings (Coombes et al. 1999). Eroksuz and Raham have found that large rainwater tanks with a capacity of up to 70 m^3 can provide 50% of the water demand for toilet flushing, washing, and watering greens in large residential buildings (Eroksuz and Raham 2010). However, Abdulla and Al.-Shareef found that RWH systems could save only from 0.27 to 19.7% of the demand for water in residential buildings located in Jordan (Abdulla and Al.-Shareef 2009). In turn, Ghisi and de Oliveira performed simulation studies of the use of rainwater for washing and toilet flushing in residential buildings in Southern Brazil. They took into account two houses, where rainwater tanks with a capacity of 3000 and 5000 L were installed. For each of them, tap water saving was 33.6% and 35.5%, respectively (Ghisi and de Oliveira 2007). However, when a larger number of houses, 62, was considered, rainwater was able to cover 34–92%. Water saving was dependent on the demand for drinking water in these buildings (Ghisi et al. 2006).

Another way to reduce tap water consumption is to use gray water in buildings (Hyde 2013). The use of gray water recycling systems (GWRS) offers significant potential for reducing the load on wastewater treatment plants and reducing the costs of tap water supply, making these systems perceived as one of the basic elements of sustainable water management (Marleni et al. 2015; Jamrah et al. 2007).

According to the European standard EN 12056-1:2000, gray water is used water free of feces and urine. They arise every day from devices such as showers, washbasins, and washing machines, and their composition is significantly different from the composition of black wastewater from the toilet bowl flushing (Marleni et al. 2015). However, gray water contains a significant content of organic compounds (Antonopoulou et al. 2013; Santasmasas et al. 2013), which makes it necessary to use pretreatment systems before using them (Grčić et al. 2015). Considering the high content of organic substances in the wastewater generated in the kitchen (Oron et al. 2014) and the significant content of detergents, bleaches, and even pathogenic organisms in wastewater from washing clothes (Maimon et al. 2014), the most commonly used gray water is discharged from washbasins, showers, and bathtubs. Treated gray water is most often used for toilet flushing and watering the garden (Penn et al. 2013a, b; Muthukumaran et al. 2011), sometimes also in hybrid systems including rainwater (Wanjiru and Xia 2018; Marinoski et al. 2018; Oviedo-Ocaña et al. 2018). Gray water recycling systems are implemented in residential buildings (Jeong et al. 2018; Lam et al. 2017), hotels (March et al. 2004), schools and universities (Laaffat et al. 2019; Shamabadi et al. 2015), airports (Couto et al. 2015), and office buildings (Hendrickson et al. 2015).

As in the case of rainwater harvesting systems, water saving as a result of using GWRS is influenced by many factors, including the type of building, water demand, amount of gray water generated in the building and habits of installation users. Jeong et al. (2018) stated that the reuse of gray water in a single-family building located in Georgia allowed reducing non-potable water demand by 17–49%, and in multifamily housing from 6 to 32%. In turn, Mourad et al. (2011) found that using GWRS to flush toilets in a single-family building in Syria can save 35% of drinking water (Mourad et al. 2011).

When analyzing the literature on rainwater harvesting systems and gray water recycling systems located in European countries, it was found that this topic had received limited interest so far, especially in the case of water reuse. Despite the fact that these solutions are used more and more often, research on their effectiveness and cost-effectiveness of use is carried out on a smaller scale than it is the case in, e.g., Australia, Brazil or Japan. In Germany, systems collecting and using rainwater in residential and public buildings have been used since the 1980s. Hermann and Schmida (1999) estimate that an average household could reduce drinking water demand by 30–60% by using rainwater to flush toilet (Hermann and Schmida 1999). In Portugal, Silva et al. (2015) stated that depending on the roof area, the saving of water intended for toilet flushing, washing, and irrigation in a single-family house could be up to 95%. Water savings in the range from 40 to even 100% were possible thanks to the application of RWHS in Genoa in Italy (Palla et al. 2011). In Poland, Słyś (2009) explored the possibility of using rainwater to flush toilets and water the garden in a single-family house. The research results showed that the reduction of water consumption from the water supply network could reach a maximum level of 84% (toilet flushing) and 95% (toilet flushing and garden watering). In this case, the water saving was influenced by the size of the roof area, the number of residents and the garden area as well as the required tank capacity resulting from these parameters (Słyś 2009). Fewkes (1999) has monitored, for a period of 12 months, a system collecting rainwater in a reservoir with a capacity of 2000 dm^3 and supplying water for toilet flushing in a house in Nottingham, Great Britain. Depending on the season, economies in water consumption ranged from 4 to 100%. In turn, in Sweden, Villarreal and Dixon (2005) investigated the water savings potential of RWHS from roof areas and noted that a drinking water saving of 30% can be achieved.

To a lesser extent than in the case of RWHS, literature with research results is available for gray water recycling systems located in Europe. Bonoli et al. (2019) conducted a Life Cycle Analysis for GWRS located in residential buildings in Bologna, Italy. They found that such analysis allowed the supporting of decision-making by comparing the environmental impacts of different technological combinations at the planning level (Bonoli et al. 2019).

The literature also includes research results on the cost-effectiveness of implementing rainwater harvesting systems and gray water recycling systems, although researchers focus more on the efficiency of water saving rather than the financial aspects of using these systems. Many researchers say that the use of RWHS and GWRS is still limited due to economic reasons, e.g., the long payback period, which in the case of small systems in single-family buildings is often several dozen years. For example, despite the fact that using treated gray water for toilet flushing would save about 35% of the potable water in Syria to a payback period of GWRS was 52 years (Mourad et al. 2011). In turn, Chilton et al. (2000) studied a rainwater harvesting system in a commercial building with a large roof area and obtained a payback period of 12 years. Cost payback periods of implementing rainwater harvesting systems were found to be longer than the lifetime of the dormitory (Devkota et al. 2015). Similar studies conducted for dorms located in Slovakia and Poland have shown that the use of RWHS is profitable only for the first location, but the discounted payback period was about 21 years (Stec and Zeleňáková 2019). The same payback period was obtained for rainwater harvesting system in a residential apartment in Nigeria (Ladc and Oloke 2015). Most research mainly focused on rainwater harvesting systems or graywater recycling systems, however, Kim et al. (2007) argued that the combination of these two systems could bring more benefits than for each separately. Ghisi and de Oliveira (2007) explored the possibilities of using these systems in Brazil as separate solutions as well as in a hybrid system. It turned out that despite the fact that these three scenarios had great potential for saving water, they were characterized by a high payback period of more than 20 years. Matos et al. Conducted a broader analysis based on three financial indicators (2015) for a commercial building located in Portugal. The economic assessment of the RWHS was made using the calculations of the net present value (NPV), the payback period (PB) and the internal rate of return (IRR). The results of this research indicate that RWHS scenarios proposed are cost-efficient and the payback period ranges from 2 to 6 years and the IRR would range from 23 to 76% (Matos et al. 2015). Wanjiru and Xia (2018) conducted a Life-Cycle Cost analysis of RWHS and GWHS located in South Africa and it turned out that high investments and low water prices meant that the implementation of these two alternative sources in the analyzed building was not financially profitable. Adewumi et al. (2010) obtained similar research results and came to the conclusion that low water charges significantly influenced end users' willingness to water recycling, and the situation could be improved by government subsidies, which would encourage saving water in this way. Similar conclusions were made by Gómez et al. (2017), who determined the financial effectiveness of RWHS for 12 single-family houses of different constructions located in Brazil. They claimed that a combination of rising water prices and reduced implementation costs improved the economic feasibility of this system. Friedler and Hadari (2006) examined the financial efficiency of the graywater recycling system in multifamily buildings, which according to their research depends on the type of system and the number of apartments. They concluded that the rotating biological reactor (RBC) graywater treatment system and the membrane bioreactor (MBR) can be economically feasible if the building has 7 stories and 40 stories, respectively.

Many researchers emphasize that public awareness and acceptance, and appropriate government policy (Pannell 2008) are very important on the way of implementing rainwater harvesting systems and graywater recycling systems. An example of this could be Brazilian Semiarid Articulation which has developed one of the world's largest social programs for the use of rainwater, with half a million cisterns built by 2016 (Gómez et al. 2017). There are also numerous examples of legislation that regulate and encourage the use of rainwater and gray water as alternative water sources. Economic incentives have been introduced in the United States, Australia, or Germany to encourage residents to implement rainwater harvesting systems (Partzsch 2009; Siems and Sahin 2015). Rahman et al. (2012) found that covering part of the initial capital expenditure related to the implementation of RWHS made certain system configurations feasible. Domènech and Saurí (2011) stated that subsidies not only making the rainwater harvesting system profitable but also can increase residents' environmental awareness.

The use of RWHS and GWRS allows not only to reduce the consumption of tap water but also can have a positive effect on the functioning of sewer systems (Gold et al. 2010; Tavakol-Davani et al. 2015; Campisano et al. 2017). The growth in the area of urbanized areas increases the sealing of the surface, which in turn intensifies the flow of rainwater discharged into sewage systems. Therefore, RWHS can not only be an alternative source of water in these areas, but also significantly reduce the outflow of rainwater from roofs to the sewer system, and thereby relieve the sewer network and storm overflows, as well as reduce the occurrence of urban floods (Sample and Liu 2014; Basinger et al. 2010; Mahmoud et al. 2014), while Schmack et al. (2019) argue that alternative water technologies, such as gray and rainwater collection systems, can reduce pressure on shallow groundwater resources, reduce the need to develop infrastructure and reduce the costs of maintaining existing water systems Penn et al. (2013a, b) came to the conclusion that graywater recycling systems can reduce sewage outflow to the sewage system and lower their flow velocity in the channels, but present only trivial impacts on the sizes of sewer pipes.

The review of previous research in the field of using rainwater and gray water as alternative water sources in buildings has shown that this is a complex issue and depends on many parameters. The most important factors that affect the hydraulic and financial efficiency of these solutions include: building location, climatic conditions, water demand, water price, and technical parameters of RWHS and GWRS. In addition, it has been observed that most research results published apply to countries outside of Europe, despite the fact that some European countries are struggling with water shortages, and for some, predicted climate change (Pavlik et al. 2014) may contribute to the appearance of water deficits in the future and deterioration in the quality of life of future generations. The average annual total amount of renewable water resources in Europe is estimated at around 6879 m^3/inhabitant (EEA 2017). This is the average value, inflated by countries with large water resources. A significant part of European countries have resources lower than 3000 m^3/year/inhabitant. For instance, Polish water resources are around 1600 m^3/person/year (Walczykiewicz 2014) and are rated as one of the lowest in Europe. Countries such as the Czech Republic, Hungary, Cyprus, and Malta have even lower resources than Poland. Water resources in Italy, Spain, France, and Belgium are estimated at about 2500 m^3/person/year, which is also well below the European average (EEA 2017). Scandinavian countries are the richest in water in Europe. There are regions in Europe where there is no water shortage, but unfortunately in many places, water shortage is a problem that Europeans face periodically or permanently. For example, in the summer of 2014, around 13% of the total European population (86 million people) lived under water scarcity (EEA 2017).

The European Union Water Framework Directive of 2000 ensures full integration of the economic and ecological aspects in the field of water quality and quantity management, i.e., in accordance with the principle of sustainable development. According to the European Environment Agency, an additional impulse to protect water resources is the observed variability of water availability (rainfall) and climate change in Europe (Krinner et al. 1999). The events of recent years have shown that extreme hydrological events, such as floods and droughts, can cause additional stress in the supply of water necessary for human health and life. Careful and efficient water management is, therefore, an important issue in Europe and a number of policies, strategies, and mechanisms should be used to ensure the sustainable use of water in the long term.

1.2   The Purpose and the Scope of the Work

The main purpose of the work is to analyze the possibilities of using alternative water sources in residential buildings in selected locations in Europe in technical, financial, social, and environmental aspects. The main purpose of the work is achieved by specific objectives, which include, above all:

- an analysis of technical possibilities of using rainwater and gray water in residential buildings,
- acquisition and analysis of real daily rainfall data from many years for selected locations in Europe necessary for conducting simulations of rainwater harvesting system,
- studies on the rainwater harvesting system simulation model in various configurations illustrating the use of rainwater for flushing toilets, washing, and watering the garden,
- determination of the hydraulic efficiency of the rainwater harvesting system and optimal tank volumes depending on the variable model parameters, including the number of inhabitants, the size of the drained area, the area of the watered garden, and the size of the demand for non-potable water,
- Life-Cycle Cost analysis of the adopted variants of plumbing in a single-family residential building, taking into account the traditional solution of the installation and installations additionally equipped with a rainwater harvesting system and graywater recycling system in their various configurations,
- studies on the hydrodynamic model of a real urban catchment area to determine the impact of implementing rainwater harvesting systems on an existing sewage system,
- surveys in eight selected European countries regarding public awareness and acceptance regarding the use of rainwater and gray wastewater in residential buildings,
- formulation of final conclusions and proposals for further research directions.

References

Abdulla FA, Al-Shreef AW (2009) Roof rainwater harvesting systems for household water supply in Jordan. Desalination 1:195–207

Adewumi J, Ilemobade A, Van Zyl J (2010) Treared wastewater reuse in South Africa: overview, potential and challenges. Resour Conserv Recycl 55:221–231

Amos CC, Rahman A, Mwangi Gathenya J (2018) Economic analysis of rainwater harvesting systems comparing developing and developed countries: a case study of Australia and Kenya. J Clean Prod 172:196–207

Antonopoulou G, Kirkou A, Stasinakis AS (2013) Quantitative and qualitative greywater characterization in Greek households and investigation of their treatment using physicochemical methods. Sci Total Environ 454–455:426–432

Australian cities. Res Conserv Recycl 54:1449–1452

Bashar M, Karim R, Imteaz MA (2018) Reliability and economic analysis of urban rainwater harvesting: a comparative study within six major cities of Bangladesh. Resour Conserv Recycl 133:146–154

Basinger M, Montalto F, Lall U (2010) A rainwater harvesting system reliability model based on nonparametric stochastic rainfall generator. J Hydrol 392:105–118

Bocanegra Martínez A, Ponce-Ortega JM, Nápoles-Rivera F, Serna-González M, Castro-Montoya AJ, El-Halwagi MM (2014) Optimal design of rainwater collecting systems for domestic use into a residential development. Resour Conserv Recycl 84:44–56

Bonoli A, Di Fusco E, Zanni S, Lauriola I, Ciriello V, Di Federico V (2019) Green smart technology for water (GST4Water): life cycle analysis of urban water consumption. Water 11(389). https://doi.org/10.3390/w11020389

Campisano A, Butler D, Ward S, Burns MJ, Friedler E, DeBusk K, Fisher-Jeffes LN, Ghisi E, Rahman A, Furumai H, Han M (2017) Urban rainwater harvesting systems: research, implementation and future perspectives. Water Res 115:195–209

Cheng C, Hong Y (2004) Evaluating water utilization in primary schools. Build Environ 39:837–845

Chilton J, Maidment G, Marriott D, Francis A, Tobias G (2000) Case study of a rainwater recovery system in a commercial building with a large roof. Urban Water 1(4):345–354

Coombes PJ, Argus JR, Kuczera G (1999) Figtree place: a case study in water sensitive urban development. Urban Water 1:335–343

Devkota J, Schlachter H, Apul D (2015) Life cycle based evaluation of harvested rainwater use in toilets and for irrigation. J Clean Prod 95:311–321

Directive 2000/60/EC of the European Parliament and of the Council of 23 October 2000 establishing a framework for Community action in the field of water policy. The European Parliament and the Council of the European Union, Brussels

Domènech L, Saurí D (2011) A comparative appraisal of the use of rainwater harvesting in single and multi-family buildings of the Metropolitan Area of Barcelona (Spain): social experience, drinking water savings and economic costs. J Clean Prod 19:598–608

EEA (2017) Use of freshwater resources in Europe 2002–2014. European Environment Agency. ISBN: 978-3-944280-57-8

EN 12056-1:2000. Gravity drainage systems inside buildings—Part 1: general and performance requirements

Ercin AE, Hoekstra AY (2014) Water footprint scenarios for 2050: a global analysis. Environ Int 64:71–82

Eroksuz E, Rahman A (2010) Rainwater tanks in multi-unit buildings: a case study for three

Fewkes A (1999) The use of rainwater for WC flushing: the field testing of a collection system. Build Environ 34:765–772

Fewkes A (2006) The technology, design and utility of rainwater catchment systems. In: Butler D, Memon FA (eds) Water demand management. IWA Publishing, London, UK

Friedler E, Hadari M (2006) Economic feasibility of on-site greywater reuse in multi-storey buildings. Desalination 190:221–234

Ghisi E, Ferreira D (2007) Potential for potable water savings by using rainwater and greywater in a multi-storey residential building in southern Brazil. Build Environ 42: 2512–2522

Ghisi E, Oliveira S (2007) Potential for potable water savings by combining the use of rainwater and greywater in houses in southern Brazil. Build Environ 42:1731–1742

Ghisi E, Montibeller A, Schmidt RW (2006) Potential for potable water savings by using rainwater: An analysis over 62 cities in southern Brazil. Build Environ 41:204–210

Ghisi E, Tavares DF, Rocha VL (2009) Rainwater harvesting in petrol stations in Brasília: potential for potable water savings and investment feasibility analysis. Resour Conserv Recycl 54:79–85

Gold A, Goo R, Hair L, Arazan N (2010) Rainwater harvesting: policies, programs, and practices for water supply sustainability. In: International low impact development conference. ASCE, San Francisco, CA

Gómez D, Teixeira Y, Girard L (2017) Residential rainwater harvesting: effects of incentive policies and water consumption over economic feasibility. Resour Conserv Recycl 127:56–67

Grčić I, Vrsaljko D, Katančić Z, Papić S (2015) Purification of household greywater loaded with hair colorants by solar photocatalysis using TiO_2-coated textile fibers coupled flocculation with chitosan. J Water Process Eng 5:15–27

Hendrickson TP, Nguyen M, Sukardi M, Miot A, Horvath A, Nelson K (2015) Life-cycle energy use and greenhouse gas emissions of a building-scale wastewater treatment and nonpotable reuse system. Environ Sci Technol 49:10303–10311

Hermann T, Schmida U (1999) Rainwater utilisation in Germany: efficiency, dimensioning, hydraulic and environmental Aspects. Urban Water 1:307–316

Hofman-Caris R, Bertelkamp Ch, de Waal L, van den Brand T, Hofman J, van der Aa R, van der Hoek J (2019) Rainwater harvesting for drinking water production: a sustainable and cost-effective solution in the Netherlands?. Water 11:511. https://doi.org/10.3390/w11030511

Hyde K (2013) An evaluation of the theoretical potential and practical opportunity for using recycled greywater for domestic purposes in Ghana. J Clean Prod 60:195–200

Imteaz MA, Shanableh A, Rahman A, Ahsan A (2011) Optimisation of rainwater tank design from large roofs: a case study in Melbourne. Australia Resour Conserv Recy 55:1022–1029

Imteaz MA, Adeboye OB, Rayburg S, Shanableh A (2012) Rainwater harvesting potential for southwest Nigeria using daily water balance model. Resour Conserv Recycl 62:51–55

Imteaz MA, Matos C, Shanableh A (2014) Impacts of climatic variability on rainwater tank outcomes for an inland city, Canberra. Int J Hydrol Sci Technol 4:177–191

Jamrah A, Al-Futaisi A, Prathapar S, Harrasi AA (2007) Evaluating greywater reuse potential for sustainable water resources management in Oman. Environ Monit Assess 137:315–327

Jeong H, Broesicke O, Drew B, Crittenden J (2018) Life cycle assessment of small-scale greywater reclamation systems combined with conventional centralized water systems for the City of Atlanta, Georgia. J Clean Prod 174:333–342

Jha MK, Chowdary VM, Kulkarni Y, Mal BC (2014) Rainwater harvesting planning using geospatial techniques and multicriteria decision analysis. Resour Conserv Recycl 83:96–111

Jones MP, Hunt WF (2010) Performance of rainwater harvesting systems in the southeastern United States. Resour Conserv Recycl 54:623–629

Kaposztasova D, Vranayova Z, Markovic G, Purcz P (2014) Rainwater Harvesting, Risk Assessment and Utilization in Kosice-city. Slovakia Procedia Eng 89:1500–1506

Kim RH, Lee S, Jeong J, Lee J, Kim Y (2007) Reuse of greywater and rainwater using fiber filter media and metal membrane. Desalination 202:326–332

Krinner W, Lallana C, Estrela T, Nixon S, Zabel T, Laffon L, Rees G, Cole G (1999) Sustainable water use in Europe. Part 1: sectoral use of water. EEA, Copenhagen

Laaffat J, Aziz F, Ouazzani N, Mandi L (2019) Biotechnological approach of greywater treatment and reuse for landscape irrigation in small communities. Saudi J Biolo Sci 26:83–90

Lade O, Oloke D (2015) Modelling rainwater harvesting system in Ibadan, Nigeria: application to a residential apartment. Am J Civil Eng Archit 3:86–100

Lade O, Oloke D (2017) Performance evaluation of a rainwater harvesting system: a case study of University Collage Hospital, Ibdan City, Nigeria. Curr J Appl Sci Technol 25:1–14

Lam C-M, Leng L, Chen P-C, Lee P-H, Hsu S-C (2017) Eco-efficiency analysis of non-potable water systems in domestic buildings. Appl Energy 202:293–307

Lee M, Kim M, Kim Y, Han M (2017) Consideration of rainwater quality parameters for drinking purposes: a case study in rural Vietnam. J Environ Manag 200:400–406

López-Morales C, Rodríguez-Tapia L (2019) On the economic analysis of wastewater treatment and reuse for designing strategies for water sustainability: lessons from the Mexico Valley Basin. Resour Conserv Recycl 140:1–12

Mahmoud WH, Elagib NA, Gaese H, Heinrich J (2014) Rainfall conditions and rainwater harvesting potential in the urban area of Khartoum. Resour Conserv Recycl 91:89–99

Maimon A, Friedler E, Gross A (2014) Parameters affecting greywater quality and its safety for reuse. Sci Total Environ 487:20–25

March J, Gual M, Orozco F (2004) Experiences on greywater re-use for toilet flushing in a hotel (Mallorca Island, Spain). Desalination 164:241–247

Marinoski AK, Rupp RF, Ghisi E (2018) Environmental benefit analysis of strategies for potable water savings in residential buildings. J Environ Manag 206:28–39

Marleni N, Gray S, Sharma A, Burn S, Muttil N (2015) Impact of water management practice scenarios on wastewater flow and contaminant concentration. J Environ Manag 151:461–471

Martin E, Buchberger S, Chakraborty D (2015) Reliability of harvested rainfall as an auxiliary source of non-potable Water. Procedia Eng 119:1119–1128

Matos C, Bentes I, Santos C, Imteaz M, Pereira S (2015) Economic analysis of a rainwater harvesting system in a commercial building. Water Resour Manag 29:3971–3986

Mekonnen MM, Hoekstra AY (2016) Four billion people facing severe water scarcity. Sci Adv 2. https://doi.org/10.1126/sciadv.1500323

Morales-Pinzón T, Lurueña R, Gabarrell X, Gasol CM, Rieradevall J (2014) Financial and environmental modelling of water hardness—implications for utilizing harvested rainwater in washing machines. Sci Total Environ 470–471:1257–1271

Moreira Neto R, Calijuri M, Carvalho I, Santiago A (2012) Rainwater treatment in airports using slow sand filtration followed by chlorination: efficiency and costs. Resour Conserv Recycl 65:124–129

Mourad K, Berndtsson J, Berndtsson R (2011) Potential fresh water saving using greywater in toilet flushing in Syria. J Environ Manag 92:2447–2453

Muthukumaran S, Baskaran K, Sexton R (2011) Quantification of potable water savings by residential water conservation and reuse—a case study. Resour Conserv Recycl 55:945–952

Mwenge Kahinda J, Taigbenu AE, Boroto RJ (2010) Domestic rainwater harvesting as an adaptation measure to climate change in South Africa. Phys Chem Earth 35:742–751

Oron G, Adel M, Agmon V, Friedler E, Halperin R, Leshem E, Weinberg D (2014) Greywater use in Israel and worldwide: standards and prospects. Water Res 58:92–101

Oudin L, Salavati B, Furusho-Percot C, Ribstein P, Saadi M (2018) Hydrological impacts of urbanization at the catchment scale. J Hydrol 559:774–786

Oviedo-Ocaña ER, Dominguez I, Ward S, Rivera-Sanchez ML, Zaraza-Peña JM (2018) Financial feasibility of end-user designed rainwater harvesting and greywater reuse systems for high water use households. Environ Sci Pollut Res 25(20):19200–19216

Palla A, Gnecco I, Lanza LG (2011) Non-dimensional design parameters and performance assessment of rainwater harvesting systems. J Hydrol 401(1–2):65–76

Pandey DN, Gupta AK, Anderson DM (2003) Rainwater harvesting as an adaptation to climate change. Curr Sci 85(1):46–59

Pannell D (2008) Public benefits, private benefits, and policy mechanism choice for land-use change for environmental benefits. Land Econ 84:225–240

Partzsch L (2009) Smart regulation for water innovation—the case of decentralized rainwater technology. J Clean Prod 17:985–991

Pavlik D, Söhl D, Pluntke T, Bernhofer C (2014) Climate change in the Western Bug river basin and the impact on future hydro-climatic conditions. Environ Earth Sci 72:4787–4799

Penn R, Friedler E, Ostfeld A (2013a) Multi-objective evolutionary optimization for greywater reuse in municipal sewer systems. Water Res 47:5911–5920

Penn R, Schutze M, Friedler E (2013b) Modelling the effects of on-site greywater reuse and low flush toilets on municipal sewer systems. J Environ Manag 114:72–83

Rahman A, Keane J, Imteaz MA (2012) Rainwater harvesting in greater Sydney: water savings, reliability and economic benefits. Resour Conserv Recycl 61:16–21

Şahin N, Manioğlu G (2018) Water conservation through rainwater harvesting using different building forms in different climatic regions. Sustain Cities Soc 44:367–377

Salvadore E, Bronders J, Batelaan O (2015) Hydrological modelling of urbanized catchments: a review and future directions. J Hydrol 529:62–81

Sample D, Liu J (2014) Optimizing rainwater harvesting systems for the dual purposes of water supply and runoff capture. J Clean Prod 75:174–194

Santasmasas C, Rovira M, Clarens F, Valderrama C (2013) Grey water reclamation by decentralized MBR prototype. Resour Conserv Recy 72:102–107

Schmack M, Anda M, Dallas S, Fornarelli R (2019) Urban water trading—hybrid water systems and niche opportunities in the urban water market—a literature review. Environ Technol Rev 8:65–81

Severis RM, da Silva F, Wahrlich J, Skoronski E, Simioni F (2019) Economic analysis and risk-based assessment of the financial losses of domestic rainwater harvesting systems. Resour Conserv Recycl 146:206–217

Shamabadi N, Bakhtiari H, Kochakian N, Farahani M (2015) The investigation and designing of an onsite grey water treatment systems at Hazrat-e-Masoumeh University, Qom, IRAN. Energy Procedia 74:1337–1346

Siems R, Sahin O (2015) Energy intensity of residential rainwater tank systems: exploring the economic and environmental impacts. J Clean Prod 113:251–262

Silva C, Sousa V, Carvalho NV (2015) Evaluation of rainwater harvesting in Portugal: application to single-family residences. Resour Conserv Recy 94:21–34

Słyś D (2009) Potential of rainwater utilization in residential housing in Poland. Water Environ J 23:318–325

Słyś D, Stec A (2014) The analysis of variants of water supply systems in multi-family residential building. Ecol Chem Eng S 21(4):623–635

Słyś D, Stec A, Zelenakova M (2012) A LCC analysis of rainwater management variants. Ecol Chem Eng S 19:359–372

Stec A, Zeleňáková M (2019) An analysis of the effectiveness of two rainwater harvesting systems located in central eastern Europe. Water 11:458. https://doi.org/10.3390/w11030458

Tavakol-Davani H, Burian SJ, Devkota J, Apul D (2015) Performance and cost-based comparison of green and gray infrastructure to control combined sewer overflows. J Sustain Water Built Environ 2:04015009

Unami K, Mohawesh O, Sharifi E, Takeuchi J, Fujihara M (2015) Stochastic modelling and control of rainwater harvesting systems for irrigation during dry spells. J Clean Prod 88:185–195

United Nations, Department of Economic and Social Affairs, Population Division (2014) World urbanization prospects: the 2014 revision, highlights

UN (2017) World population prospects: the 2017 revision. Population division. ESA/P/WP/248; United Nations. Department of Economic and Social Aairs, New York, NY, USA

Urbaniec K, Mikulčić H, Rosen MA, Duić N (2017) A holistic approach to sustainable development of energy, water and environment system. J Clean Prod 155:1–11

Villarreal EL, Dixon A (2005) Analysis of a rainwater collection system for domestic water supply in Ringdansen, Norrkoping, Sweden. Build Environ 40:1174–1184

Vörösmarty CJ, Green P, Salisbury J, Lammers RB (2000) Global water resources: vulnerability from climate change and population growth. Science 289:284–288

Walczykiewicz T (2014) Scenarios of Water Resources Development in Poland up to 2030. Water Resour 41:763–773

Wanjiru E, Xia X (2018) Sustainable energy-water management for residential houses with optimal integrated grey and rain water recycling. J Clean Prod 170:1151–1166

Ward S, Memon FA, Butler D (2012) Performance of a large building rainwater harvesting system. Water Res 46:5127–5134

WWAP United Nations World Water Assessment Programme (2015) The United Nations World
Water Development Report 2015: Water for a Sustainable World. Paris, UNESCO

Yana X, Ward S, Butler D, Daly B (2018) Performance assessment and life cycle analysis of
potable water production from harvested rainwater by a decentralized system. J Clean Prod
17:22167–22173

Zaizen M, Urakawa T, Matsumoto Y, Takai H (1999) The collection of rainwater from dome
stadiums in Japan. Urban Water 1:355–359

Zhang X, Hu M (2014) Effectiveness of rainwater harvesting in runoff volume reduction in a planned
industrial park. China Water Resour Manag 28:671–682

Zhang Y, Grant A, Sharma A, Chen D, Chen L (2010) Alternative water resources for rural residential
development In Western Australia. Water Resour Manag 24:25–36

Zhang S, Zhang J, Yue T, Jing X (2019) Impacts of climate change on urban rainwater harvesting
systems. Sci Total Environ 665:262–274

Chapter 2
Water Resources

Abstract Water resources are particularly important among other natural resources available on the globe. Water is the most widespread substance that plays a very important role both for the environment and in human life. It is necessary for all known living organisms to function properly. The most important are the resources of fresh water, which for centuries people have been using for various purposes. For hundreds of years, the human impact on these resources was small and had a more local character. The natural properties of water resources enabling their self-purification and renewal have allowed for a long time keeping them in good quality and quantity. This gave birth to the conviction that water resources were unchanged and not inexhaustible, which in turn caused excessive and unlimited exploitation in combination with minimal activities aimed at protecting their quality (Shiklomanov 1998).

The condition of water resources has changed drastically over the past several decades. In many regions of the world, adverse effects of long-term, often irrational, human activities have been noticed. This applies to the direct use of water resources and surface transformations in many catchments. This is mainly caused by an increase in the number of population in the world, urbanization, increased use of water for industrial, agricultural and energy purposes, and the construction of numerous water reservoirs on all continents. This has significantly affected the natural hydrological cycle and the quality and quantity of water resources available. The spatial and temporal distribution thereof depends not only on the climate but also on the economic activity of people in a given region. Currently, in many places in the world water resources are so impoverished and polluted that they can not cover the ever-increasing demand for water.

Many definitions of water resources are used. According to the International Hydrological Dictionary (2012), water resources are available or those that can be available for use in a given region, with a specified quantity and quality, during a given

© Springer Nature Switzerland AG 2020

A. Stec, *Sustainable Water Management in Buildings*,

Water Science and Technology Library 90,

https://doi.org/10.1007/978-3-030-35959-1_2

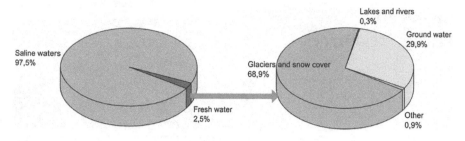

Fig. 2.1 World water resources (based on Shiklomanov 1998)

period, with specific needs. Taking into account the states of water concentration occurring in nature, water resources can be divided into:

- saltwater in the oceans and seas,
- fresh water found in rivers, lakes, and other surface reservoirs,
- water contained in glaciers and ice sheets,
- water occurring in the form of clouds and water vapor,
- water in the upper layers of the earth's crust and soil cover.

Estimation of the total amount of water occurring on the Earth is complicated since water is very dynamic and undergoes constant changes from liquid to solid and gaseous and vice versa. Typically, the amount of water available in the hydrosphere is determined. It is all free water occurring in the liquid, solid, or gaseous state in the atmosphere, on the surface of the Earth and in the Earth's crust to a depth of 2000 m (UN 2012; Shiklomanov 1998). It is assumed that there are almost 1400 million cubic kilometers of water in the Earth's hydrosphere. However, a huge amount (97.5%) is salty water, and only 2.5% is fresh water that can be a source of a potable water (Fig. 2.1) (Shiklomanov 1998). Most of this fresh water (68.7%) occurs in the form of ice and snow cover in Antarctica, the Arctic, and in mountainous regions. Almost 30% of freshwater resources are groundwater, and only 0.3% occur in lakes, rivers and surface reservoirs, from where they are most easily collected for the needs of human activity.

The values shown in Fig. 2.1 are static ones determined for a longer period of time. In shorter time intervals (year, season, and month) the volume of water stored in the hydrosphere is constantly changing as a result of its exchange between oceans and seas, lands, and atmosphere. This process is called the hydrological cycle and is schematically shown in Fig. 2.2.

The hydrological cycle of the water cycle in nature consists of many phenomena, among which one can be distinguished (Słyś 2013):

- evaporation from the surface of oceans and seas,
- evaporation of water from the surface of lakes and rivers,
- evapotranspiration,
- atmospheric precipitation,
- sublimation of snow and ice,

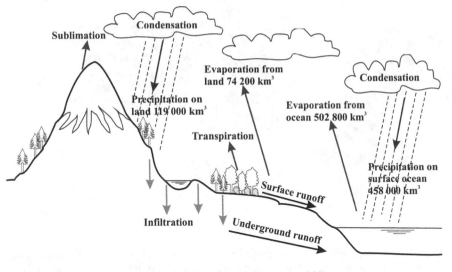

Fig. 2.2 Water cycle in the environment (based on Shiklomanov 1998)

- infiltration of water to the ground,
- surface runoff,
- subsurface runoff,
- interception.

Solar heat evaporates water from the land, oceans, seas, lakes, and rivers, and causes that it moves to the atmosphere, from which water returns again in the form of precipitation. This precipitation is the main source of water occurring on land in rivers and lakes. Part of the precipitation evaporates, part becomes leaky and supplies groundwater. The remaining part flows down and through the rivers returns to the seas and oceans, from where it evaporates into the atmosphere. This process is repeated many times.

The water cycle, mainly involving atmospheric precipitation and the process of evaporation from various surfaces, occurring locally is called a small circulation of water. The large water cycle consists of the continental and atmospheric phases, and its stages are also the flow to the oceans and evaporation from their surface.

When analyzing quantitatively the components of the annual global hydrological cycle (Fig. 2.2), it can be seen that the largest part is water which evaporates from the surface of the oceans and seas (502,800 km^3 of water) and land (74,200 km^3 of water). Therefore, the annual water circulation on the Earth is 777,000 km^3. The same amount of water returns in the form of precipitation to the surface of the oceans in the amount of 458,000 km^3 and land surface at the level of 119,000 km^3. The difference between precipitation and evaporation is determined by the total annual runoff of water from the land surface to the oceans, which includes the runoffs to rivers and the underground ones (Shiklomanov 1998). These runoffs are the main

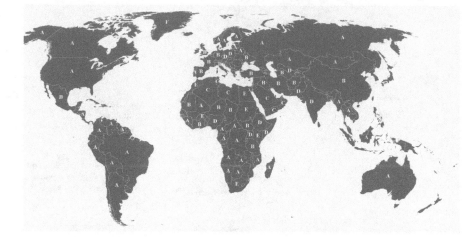

Fig. 2.3 Total renewable water resources per inhabitant, >5000 m^3/year—A (water stress possible locally), 1700–5000 m^3/year—B (occasional or local water stress), 1000–1700 m^3/year—D (water stress), 500–1000 m^3/year—E (chronic water scarcity), <500 m^3/year—C (absolute water scarcity), No data—F (based on AQUASTAT 2015)

sources of fresh water that provide coverage for life needs and those resulting from human activities.

Potable water resources are renewable, and this process is regulated by hydrological cycles. However, their distribution on the globe is very uneven and does not coincide in many regions with the demand for water. Almost 70% of the world's freshwater resources are concentrated in glaciers and ice sheets occurring mainly in Greenland and Antarctica, so places virtually inaccessible to people. The specific water availability reflects the value of the actual renewable water resources per capita. Figure 2.3 shows the distribution of global water resources in relation to one inhabitant. On the basis of this map, it is possible to indicate the regions where water shortages can occur with varying frequencies. Water stress is often defined as the ratio of water use to water availability. It is assumed that water availability below 1000 m^3/capita/year causes water stress in a given area (UN 2012). Physical water scarcity occurs when there is not enough water to meet all demands (Vanham et al. 2018). Recently, 4 billion people face severe water stress during at least 1 month per year, and 1.8 billion at least 6 months per year (Mekonnen and Hoekstra 2016). This includes populations throughout northern Africa, the Mediterranean region, the Near East, the Middle East, Southern Asia, Mexico, and South Africa (UN 2012). According to many forecasts, the situation in the world with the availability of safe water will deteriorate. This will be influenced by population growth, climate change (a decrease in rainfall, long-term drought), growing environmental degradation, and poor management of available water resources.

The average value of renewable global water resources of 42,700 km^3 per year is not only unevenly distributed in the world but also very variable over time. The largest water resources occur in Asia and South America (13,500 and 12,000 km^3

annually). The smallest are available in Europe (2900 km^3 per year) and Australia (2400 km^3 per year) (Shiklomanov 1998). When analyzing the availability of water in individual countries, it can be seen that its largest resources are concentrated mainly in six countries in the world: Brazil, Russia, Canada, the United States, China, and India (Shiklomanov 1998). The variability of the availability of water resources over time can range from 15 to 25% of their average values.

In Europe, about 4000 km^3 of water comes from precipitation during the year, which is 4.1% on a global scale (FAO 2014). Over half of the rainfall (52%) in Europe returns to the atmosphere through evapotranspiration, and 13% is used by ecosystems. About 11% of this rainfall becomes filtered into the ground, feeding groundwater resources. The remaining part of the rainfall, at the level of 24%, forms surface runoff and also meets the direct demand of ecosystems for water (EEA 2017a, b). In the scale of a given country, the amount of water resources, apart from precipitation, is also influenced by external inflows from areas located above. Since many rivers in Europe are cross-border waterways, external inflows to some countries are of great importance. Water resources in six European countries are more than 50% dependent on the amount of water coming from other countries. These include: Serbia, Hungary, the Netherlands, Bulgaria, Slovakia, and Estonia. In the case of Serbia, Hungary, and the Netherlands, this level exceeds 90%. The water availability of Latvia, Lithuania, Germany, Slovenia, and Albania depends on 30–50% of external inflows. To a lesser extent, countries such as Poland, Sweden, Ireland, France, the Czech Republic, Finland, Turkey, and Romania depend on them. In addition, Malta and Cyprus, which are located on islands, do not receive water from neighboring catchments (EEA 2017a, b).

The annual total amount of renewable water resources in Europe is estimated at around 6879 m^3 per inhabitant (EEA 2017a, b). However, this is an average value inflated by countries with rich water resources. When analyzing the values shown in Fig. 2.4, it can be seen that a significant part of European countries has resources below 3000 m^3/year/inhabitant. There are regions in Europe where there are no water shortages, but unfortunately in many places, water scarcity is a fact that Europeans are grappling with periodically or permanently. For instance, in the summer of 2014, around 86 million inhabitants lived under water scarcity, which corresponds to almost 13% of the total European population. Water shortages are felt by the inhabitants of Europe not only in the summer but also in spring and winter, when respectively, 11.4% and 6.6% of the total European population lived under water stress (EEA 2017a, b).

Over the past nearly 60 years, global renewable freshwater resources have dropped drastically from 13,200 m^3/year/inhabitant to 6,879 m^3/year/inhabitant (World Bank 2019). Similar changes are also observed in Europe. Renewable freshwater resources, especially in Western and Southern Europe, have decreased by around 24% during this period. The reasons for these significant changes are seen mainly in an increase of population in Europe and climate change, especially those taking place in the south of Europe. The European population has grown by about 117 million in the past three decades (1987–2015), which also contributed to the intensification of the

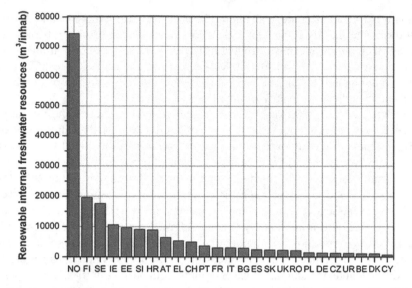

Fig. 2.4 Renewable freshwater resources in European countries, NO—Norway, FI—Finland, SE—Sweden, EE—Estonia, SI—Slovenia, HR—Croatia, CH—Switzerland, AT—Austria, PT—Portugal, FR—France, IT—Italy, BG—Bulgaria, ES—Spain, SK—Slovakia, UK—The United Kingdom, RO—Romania, PL—Poland, DE—Germany, CZ—The Czech Republic, UR—Ukraine, BE—Belgium, DK—Denmark, CY—Cyprus, HU—Hungary, MT—Malta (based on AQUASTAT 2015)

urbanization process. In the same period, the number of people living in urban areas increased by around 120 million (EEA 2017a, b).

As meteorological observations show, in parallel to global warming, the average annual temperature in Europe increased by 1.45–1.59 °C compared to the preindustrial period (EEA 2017a, b). In addition to changes in the temperature, some significant ones in precipitation tendencies are also noticeable. Since 1960, in some parts of Europe, for example, in Portugal, there has been a 90 mm drop in precipitation per decade, in particular during the summer. Atmospheric precipitation has decreased by about 20 mm/decade in most Southern European countries (EEA 2017a). Climate change in this area also causes changes in long-term river flows in various parts of Europe. The flow of rivers decreases especially during the summer months, in which there is a high demand for water.

Numerous forecasts predict that these unfavorable trends will continue and deepen in the coming years (EEA 2017a, b). Therefore, in many European countries, appropriate technical and legal measures have been taken to protect water resources. The European Union's Water Framework Directive (WFD 2000) obliges the Member States to promote the sustainable use of water resources based on long-term conservation and to ensure a balance between the collection and discharge of surface and groundwater drainage into the environment. These activities should focus on the quantitative and qualitative protection of water resources through appropriate flood protection and spatial development strategies for urban areas, design of ecological technical infrastructure, application of water-saving technologies, as well as acquiring water from alternative sources and its recycling.

References

AQUASTAT (2015) fao.org/nr/water/aquastat/data/query/index.html?lang=en

EEA (2017a) Mean precipitation. http://www.eea.europa.eu/data-and-maps/indicators/european-precipitation-1/assessment-1. Accessed Mar 2019

EEA (2017b) Use of freshwater resources in Europe 2002–2014. European Environment Agency. ISBN: 978-3-944280-57-8

FAO (2014) Freshwater availability—precipitation and internal renewable water resources (IRWR), AQUASTAT

International Glossary of Hydrology (2012) United Nations Educational, Scientific and Cultural Organization. ISBN 978-92-3-001154-3

Mekonnen MM, Hoekstra AY (2016) Four billion people facing severe water scarcity. Sci Adv 2(2). https://doi.org/10.1126/sciadv.1500323

Shiklomanov I (1998) World water resources. A new appraisal and assessment for the 21st century. United Nations Educational, Scientific and Cultural Organization, France

UN (2012) Global water futures 2050. United Nations Educational, Scientific and Cultural Organization, France

Vanham D, Hoekstra AY, Wada Y, Bouraoui F, de Roo A, Mekonnen MM, an de Bund WJ, Batelaan O, Pavelic P, Bastiaanssen WGM, Kummu M, Rockström J, Lium J, Bisselink B, Ronco P, Pistocchi A, Bidoglio G (2018) Physical water scarcity metrics for monitoring progress towards SDG target 6.4: an evaluation of indicator 6.4.2 "Level of water stress". Sci Total Environ 613–614:218–232

WFD (2000) Directive 2000/60/EC of the European Parliament and of the Council of 23 October 2000 establishing a framework for community action in the field of water policy

World Bank (2019) Renewable internal freshwater resources per capita. http://data.worldbank.org/indicator/ER.H2O.INTR.PC. Accessed Mar 2019

Chapter 3
Demand for Water in the Building

Abstract Over the years, water consumption, its structure, and size have undergone constant changes. Since the commissioning of the first water supply systems, water intake has been systematically growing until the 1980s. Such high consumption of water resulted mainly from the lack of its measurement, as well as low environmental awareness of recipients. The further development of water supply systems in the 1990s was characterized by smaller increases in water abstraction rates. This was mainly influenced by the development of installation techniques, an increase in water prices, introduction of the obligation to measure water consumption, and changing the habits of users of water supply installations. Considering the current and forecasted limitations of accessibility to water sources and the resulting need to save it, it should be anticipated that further development of installation technology will focus on seeking technical solutions that not only ensure high comfort of use but also contribute to a significant reduction in water consumption. This will force the need for alternative sources of water, such as rainwater and gray sewage, which will partly replace tap water.

Water in households is used for catering, body washing, washing, toilet bowls flushing, performing cleaning work inside and outside, and for watering greenery. The daily consumption depends mainly on the climate, water resources, individual habits of users and their age, the standard of building sanitary equipment and water price (Dias et al. 2018; Hussien et al. 2016; Justes et al. 2014; Beal et al. 2013; Butler and Memon 2006). The main task of the water supplied to the building is to meet the basic needs of its residents. According to some researchers, this is a value of 50 L/inhabitant/day, which should satisfy four basic human needs, as water: for drinking, personal hygiene, preparation of modest meals, and for sanitary services (Gleick 1996). However, in most countries of the world, daily water consumption in households per person is much higher, reaching 335 L in Canada or as much as 380 L in the US (Ramulongo et al. 2017). Considering that people use water for many different activities, and some of them are more important than others, the World Health Organization has set a hierarchy of water consumption for individual purposes (Fig. 3.1). It shows that one needs only 10–20 L of water a day to survive,

© Springer Nature Switzerland AG 2020

A. Stec, *Sustainable Water Management in Buildings*,
Water Science and Technology Library 90,
https://doi.org/10.1007/978-3-030-35959-1_3

Fig. 3.1 Hierarchy of water requirements (based on WHO 2013)

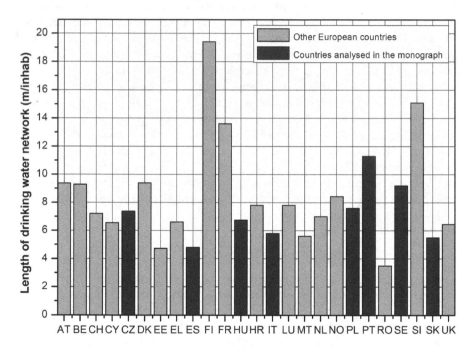

Fig. 3.2 Length of the potable water network per connected inhabitant, AT—Austria, BE—Belgium, CH—Switzerland, CY—Cyprus, CZ—The Czech Republic, DK—Denmark, EE—Estonia, EL—Greece, ES—Spain, FI—Finland, FR—France, HU—Hungary, HR—Croatia, IT—Italy, LU—Luxembourg, MT—Malta, NL—The Netherlands, NO—Norway, PL—Poland, PT—Portugal, RO—Romania, SE—Sweden, SI—Slovenia, SK—Slovakia, UK—The United Kingdom (based on EurEau 2017)

but this only applies to a short period of time and exceptional situations. In order to maintain adequate hygiene, approximately 70 L/inhabitant/day water is needed for bathing, washing, and cleaning at home (WHO 2013).

Currently, in Europe, the main water source for buildings is the water supply network, and its total length is 4 225 527 km (EurEau 2017). Figure 3.2 shows the

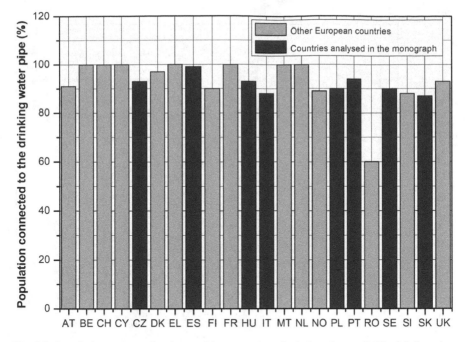

Fig. 3.3 Population connected to the potable water network, designations as in Fig. 3.2 (based on EurEau 2017)

average length of the water supply network per user with the division into countries analyzed in this monograph and other European countries for which data was available. This length ranges from 3.5 m/inhabitant (Romania) to 19.6 m/inhabitant (Finland). The significant differences observed result from the population density in each country. 95% of Europe's residents are supplied with clean, safe, and drinking water. Detailed data in this respect for selected countries is presented in Fig. 3.3. For the majority of the analyzed regions, the connection factor for the water supply network is over 90%.

Water from the water supply network is supplied to residential, service, and industrial buildings. As the data presented in Fig. 3.4 shows in all countries, the largest water consumption is observed in households and constitutes from 65 to 80% of the total water sales from the network (EurEau 2017). Therefore, it is very important from the point of view of water resources protection to limit the use of water supplied from the water supply network in residential buildings. The amount of water that fully satisfies all human needs during the day ranges from about 100 to 150 dm^3. However, this value also depends on the country (Fig. 3.5). The data presented in Fig. 3.5 were determined on the basis of statistical data, which is why significant differences in daily water consumption by residents of different countries are observed. The average value in Europe is 128 L/inhabitant/day (EurEau 2017). In some countries, such as Cyprus, Greece, Spain, Italy, and Portugal, daily household water consumption far exceeds the European average, reaching even 245 L/inhabitant/day. This trend

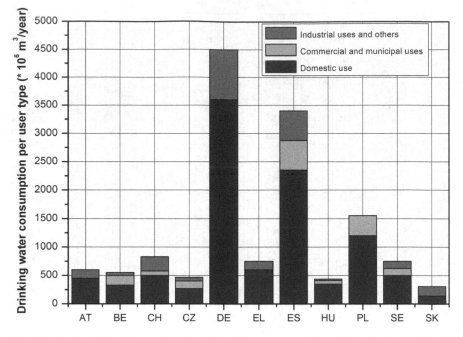

Fig. 3.4 Potable water consumption for selected European countries, designations as in Fig. 3.2 (based on EurEau 2017)

is also noticeable when analyzing annual water consumption in households, which depends on the number of people living in a given farm. The average household size in Europe ranges from 1.75 for Portugal to 2.75 persons for Poland and Ireland (Fig. 3.6). The European average is 2.3 people (EurEau 2017).

The number of people in the household determines the annual consumption of water, which, depending on the country, varies in a wide range from 80 m^3 to even 220 m^3. These data are presented in Fig. 3.7. The annual average water consumption of a household in Europe is 112 m^3 (EurEau 2017).

The amount of water used in residential buildings is affected not only by the climate, availability of water resources, or the number of inhabitants, but also the type of household (a flat, a detached house, a semi-detached house) and its size, age of people, and the season (Arbués et al. 2010; Russel and Fielding 2010; Butler and Memon 2006). The number of people living in the household has the direct impact on water consumption and it is obvious that with the growth in the number of inhabitants, the total water demand increases, but at the same time, the dependence of the decrease in water consumption per person is observed (Butler 1991). For instance, in a single-person household, the water consumption per person is 40% higher than in two-person and 73% higher than in a four-person household (Butler and Memon 2006). Russac et al. (1991) observed that the demand for water is higher in detached houses than in flats, which may be due to the use of water for watering the garden or larger household area.

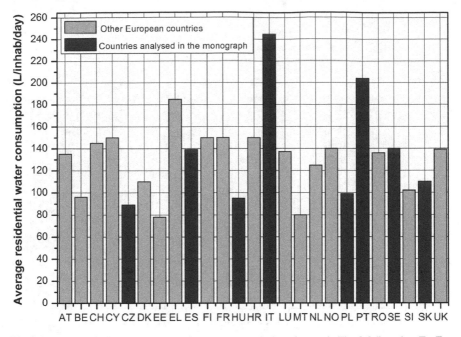

Fig. 3.5 Average daily water consumption per person, designations as in Fig. 3.2 (based on EurEau 2017)

When analyzing the structure of water consumption as "End use", it can be noticed that over a day, man uses more than two-thirds of water for hygienic purposes using it for bathing, washing hands, and toilets flushing (Energy Saving Trust 2013). However, the final water consumption for individual purposes depends on the country, people's behavior and habits, and the type and technical condition of sanitary facilities. As a consequence, in many countries, detailed data in this area are not available or appear in a limited form as local technical reports and case studies. Noteworthy is the example of Great Britain, where since 2010 the online tool Water Energy Calculator is being operated, which allows residents to determine the detailed water consumption at home with a breakdown into goals (Energy Saving Trust 2013). Thanks to this information, they are able to take action to reduce water consumption and thereby reduce water bills by replacing water-saving devices. According to data obtained from this calculator, water consumption by use in the United Kingdom is in the ranges shown in Fig. 3.8. Currently, the largest share is the consumption of water for washing the body 33% (shower 25%, bath 8%), and toilet flushing 22%. In recent years, this hierarchy has been reversed. According to the European Environment Agency, the largest amount of water by 2000 was used in the UK to flush toilets (EEA 2001). The situation was similar in many other European countries, where 25–30% of the total water consumption in households was consumed for this purpose (EEA 2001).

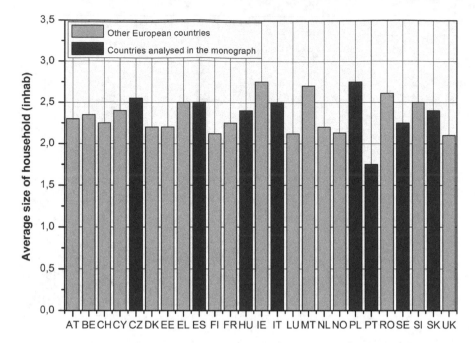

Fig. 3.6 Average household size in Europe, designations as in Fig. 3.2 (based on EurEau 2017)

As in the United Kingdom, in the Netherlands, the main use of household water is shower (40%) and toilet (28%) (Dutch 2017). Water for other purposes accounts for a total of 32% of the total amount of water consumed by a person during the day (Fig. 3.9). Also in Portugal (Fig. 3.10), household water is mainly used for residential hygiene (36%) and for toilets flushing (21%). Proportions in this respect are similar to the daily consumption of water in the Netherlands. A slightly changed hierarchy in the structure of water consumption in homes is observed, inter alia, in the United States and Poland, where the most water is used to flush toilets, and the second place is occupied by the consumption of water for body washing. However, the differences between the two end uses are small, at the level of several percents. The average daily water consumption divided into consumption types is shown in Figs. 3.11 and 3.12 for the United States and Poland, respectively.

In addition to the consumption of water inside the building, it is also used for external work within it: watering the garden, washing the car, or washing the driveway. The demand for these purposes is different and depends on the region, climate, season, behavior of residents, and applicable legal provisions in a given country. Research in the UK shows that outdoor water use in households is less than 10% of total water use during the day (Energy Saving Trust 2013). In 42% of the households, a hosepipe is used for watering the garden, and in 48% of the cases, car washing is carried out using a bucket (Energy Saving Trust 2013).

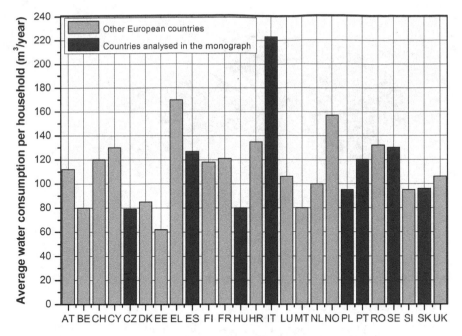

Fig. 3.7 Average annual water consumption per household, designations as in Fig. 3.2 (based on EurEau 2017)

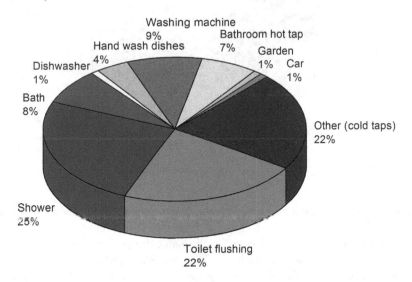

Fig. 3.8 Water consumption by use in Great Britain (based on Energy Saving Trust 2013)

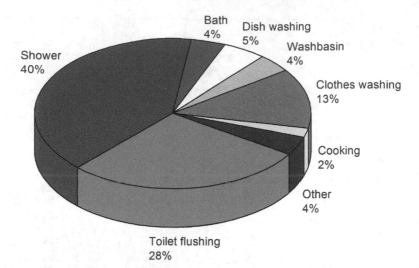

Fig. 3.9   Water consumption by use in the Netherlands (based on Dutch 2017)

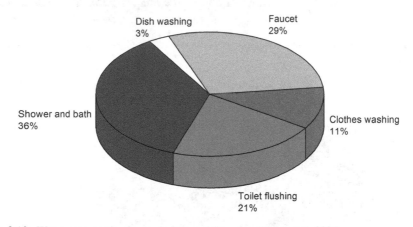

Fig. 3.10   Water consumption by use in Portugal (based on Vieira et al. 2007)

The frequency of these activities outside has a decisive influence on water consumption. For example, 22% of households analyzed in Poland water the garden during the summer once a week, while the same frequency is declared by 43% households in Greece (Shan et al. 2015). Both the Poles and the Greeks most often use hosepipe (47 and 53%), less frequently a watering can (18 and 19%), and a bucket (6 and 7%) (Shan et al. 2015). The amount of water used for watering the garden and washing the car can be: 20–30 dm^3 for washing the car with a bucket, 50–100 dm^3 for washing cars with a hose, and 10–20 dm^3/min for watering the green with a hose (Chudzicki and Sosnowski 2009).

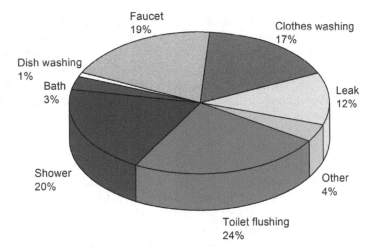

Fig. 3.11 Water consumption by use in the USA (based on WRF 2016)

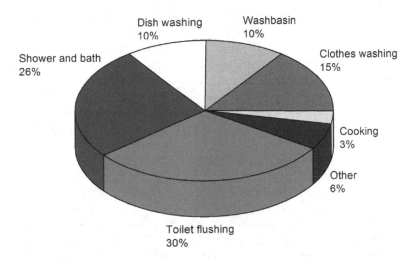

Fig. 3.12 Water consumption by use in Poland (based on Chudzicki and Sosnowski 2009)

A much greater demand for water for external purposes is observed in countries with a dry climate and regions, where long-term periods of drought often occur. For instance, in South African countries in a typical household, 35% of water is used for watering the garden and this is the largest share in the total daily water demand (CWS 2019). In turn, in the United States, with around 29 billion gallons of water used daily by households, almost 9 billion gallons (30%) is intended for outdoor water use, and in the summer months or in arid regions this share can reach up to 70% (EPA 2019). This consumption also varies depending on the season. It is noticeable not only in a temperate climate, where in the autumn and winter the gardens are not watered but also in a dry and hot climate, where there are periods with reduced demand for water for watering. In Australia, these differences are significant, from over 18% of the total daily water consumption in summer to 3% in winter (Gan and Redhead 2013).

In recent years, a decrease in water consumption in households, both those used outside and inside, has been observed in many countries. This is particularly true for regions that have hitherto had a high demand for water or areas with small water resources. A dozen or so years ago, in some regions, water intended for irrigation could account for up to 54% of daily water consumption in households, and now it is at the level of several percents (Willis et al. 2013). Such significant changes are mainly caused by the introduction of appropriate legal provisions, the use of water-saving devices, the increase in water prices, the use of alternative water sources and raising the awareness of societies (Tortajada et al. 2019; Cominola et al. 2015).

The easiest way to reduce water consumption in buildings is to implement water-saving devices, including a high-efficiency showerhead and a washing machine, and a dual flush toilet. Mayer et al. (2004) found that replacing traditional household appliances with high water efficiency could reduce water consumption in homes by almost 50%. Inman and Jeffrey (2006) came to similar conclusions. In turn, other researchers emphasized that the modernization and replacement of devices are just one of the ways to save water and this was not enough. The best results can be obtained by using high-efficiency household appliances in combination with changing the behavior of residents, e.g., turning water while brushing teeth (Mallet and Melchiori 2016) and through the use of rainwater (Willis et al. 2013).

The daily water consumption can be divided into rational consumption, i.e., that which is necessary for human functioning and consumption caused by water losses. These losses arise as a result of water leaks due to some failure in equipment and water supply fittings as well as wastage. Along with the improvement of the technical condition of water supply installations, the problem of large water leaks has been limited, but the problem of waste is still valid. This concept is relative and depends on the individual habits of users. However, when analyzing the structure of water consumption, it can be noticed that within 24 h a person uses it in more than 50% for toilets flushing, laundry, and other economic purposes. The use of very high-quality water that meets the requirements set for potable water for purposes where this quality is not required is an unreasonable waste. Therefore, in order to rationalize water consumption in households, the possibility of using water of lower quality originating from alternative sources should be considered. The use of dual installations in buildings will contribute not only to reducing the fees incurred for supplying tap water but also to the protection of natural water resources.

References

Arbués F, Villanúa I, Barberán R (2010) Household size and residential water demand: an empirical approach. Aust J Agric Resour Econ 54:61–80

Beal CD, Stewart RA, Fielding K (2013) A novel mixed method smart metering approach to reconciling differences between perceived and actual residential end use water consumption. J Clean Prod 60:116–128

Butler D (1991) A small-scale study of wastewater discharges from domestics appliances. Water and Environ J 5:178–185

Butler D, Memon FA (2006) Water demand management. IWA Publishing, London. ISBN 1483390787

Chudzicki J, Sosnowski S (2009) Instalacje wodociągowe: projektowanie, wykonanie, eksploatacja, Wydawnictwo „Seidel-Przywecki" Sp. z o.o., Warszawa

Cominola A, Giuliani M, Piga D, Castelleti A, Rizzoli A (2015) Benefits and challenges of using smart meters for advancing residential water demand modeling and management: a review. Environ Model Softw 72:198–214

CWS (2019). http://www.capewatersolutions.co.za/2010/02/06/typical-household-water-consumption/

Dias TF, Kalbusch A, Henning E (2018) Factors influencing water consumption in buildings in southern Brazil. J Clean Prod 184:160–167

Dutch (2017) Dutch drinking water statistics 2017. www.vcwin.nl

EEA (2001) Household water consumption. European Environment Agency

Energy Saving Trust (2013) At home with water. www.energysavingtrust.org.uk

EPA (2019). https://www.epa.gov/watersense/start-saving

EurEau (2017) Europe's water in figures. An overview of the European drinking water and waste water sectors. European Federation of National Associations of Water Services

Gan K, Readhead M (2013) Melbourne residential water use studies. Smart Water Fund

Gleick PH (1996) Basic water requirements for human activities. Meet Basic Needs Water Int 21:83–92

Hussien WA, Memon FA, Savic DA (2016) Assessing and modelling the influence of household characteristics on per capita water consumption. Water Resour Manag Int J Publ Eur Water Resour Assoc (EWRA) 30(9):2931–2955

Inman D, Jeffrey P (2006) A review of residential water conservation tool performance and influences on implementation effectiveness. Urban Water J 3:127–143

Justes A, Barberán R, Farizo BA (2014) Economic valuation of domestic water uses. Sci Total Environ 472:712–718

Mallett RK, Melchiori KJ (2016) Creating a water-saver self-identity reduces water use in residence halls. J Environ Psychol 47:223–229

Mayer P, DeOreo W, Towler E, Martien L, Lewis D (2004) Tampa water department residential water conservation study: the impacts of high efficiency plumbing fixture retrofits in single-family homes. Aquacraft, Inc Water Engineering and Management, Tampa

Ramulongo L, Nethengwe NS, Musyoki A (2017) The nature of urban household water demand and consumption in Makhado Local Municipality: A case study of Makhado Newtown. Procedia Environ Sci 37:182–194

Russac D, Rushton K, Simpson R (1991) Insight into domestic demand from metering trial. Water and Environ J 5:342–351

Russell S, Fielding K (2010) Water demand management research: a psychological perspective. Water Resour Res 46:W05302. https://doi.org/10.1029/2009wr008408

Shan Y, Yang L, Perren K, Zhang Y (2015) Household water consumption: Insight from a survey in Greece and Poland. Procedia Eng 119:1409–1418

Tortajada C, González-Gómez F, Biswas A, Buurman J (2019) Water demand management strategies for water-scarce cities: the case of Spain. Sustain Cities Soc 45:649–656

Vieira P, Almeida MC, Baptista JM, Ribeiro R (2007) Household water use: a Portuguese field study. Water Sci Technol: Water Supply 7(5–6):193–202

WHO (2013) Technical notes on drinking-water, sanitation and hygiene in emergencies. World Health Organization

Willis RM, Stewart RA, Giurco DP, Talebpour MR, Mousavinejad A (2013) End use water consumption in households: impact of socio-demographic factors and efficient devices. J Clean Prod 60:107–115
WRF (2016) Residential end uses of water. Executive report, Version 2. Water Research Foundation

Chapter 4
Alternative Water Resources

Abstract Water resources of the Earth are huge and would be able to satisfy the needs of all humanity. However, their uneven distribution and irrational management by a man mean that in many countries, also European ones, water supply is a very big problem. According to numerous forecasts, mainly concerning climate change, the water shortage may still get worse in the coming years. To counteract this, it is necessary, inter alia, to introduce a sustainable water management strategy that will take into account alternative sources of water. When looking for them, special attention was paid to rainwater, which is characterized by a small degree of pollution and gray water which is available in the building regardless of the climate and weather conditions.

4.1 Rainwater Harvesting

4.1.1 The Characteristics of Rainwater

Atmospheric rainwater is a product of the condensation of water vapor, which occurs mainly in the lower part of the Earth's atmosphere. Its quantity and state of aggregation depend on the temperature and humidity conditions of the air. Precipitation occurs in various forms and different intensities, and its classification has been made by the World Meteorological Organization. It proposed a division of solid and liquid precipitation, which is presented in Table 4.1.

Precipitation occurring in a given area is not only characterized by varying intensity, but also its distribution during the year is uneven and sometimes subject to considerable fluctuations in subsequent years. The annual sums of rainfall are close to the average value called the normal precipitation determined on the basis of measurement data from a period of at least 30 years. The amount of normal precipitation depends on many factors, the most important of which are: elevation of the area relative to sea level, latitude, directions of winds, occurrence in the vicinity of mountains and seas.

© Springer Nature Switzerland AG 2020
A. Stec, *Sustainable Water Management in Buildings*,
Water Science and Technology Library 90,
https://doi.org/10.1007/978-3-030-35959-1_4

Table 4.1 Classification of atmospheric precipitation (Sumner 1988)

Type of precipitation	Precipitation intensity	Precipitation characteristics
Rain	Intensive	Precipitation of at least 4.0 mm/h
	Moderate	Precipitation in intensity from 4.0 to 0.5 mm/h
	Poor	Precipitation less than 0.5 mm/h
Heavy rain	Heavy	Precipitation of at least 50.0 mm/h
	Rapid	Precipitation from 50.0 to 10.0 mm/h
	Moderate	Precipitation less than 10.0 mm/h
Drizzle	Thick	Visibly obstructing and accumulating at speeds of up to 1 mm/h
	Moderate	It causes moisture to flow on the surface of windows and roads
	Poor	Feels like the face, but does not cause drainage
Snow	Intensive	Reduces visibility to small distances and causes snow cover growth at speeds exceeding 4 cm/h
	Moderate	Consisting of large petals falling so densely that they impede visibility, snow cover growth to 4 cm/h
	Poor	The flakes are small and fall rarely, the accumulation intensity does not exceed 0.5 cm/h
Hail	Heavy	Unique, containing a large portion of hailstones with a diameter of more than 6.5 mm
	Moderate	Abundant, causing the ground to be white, and after melting it will create a significant amount of rainfall
	Poor	Rare hail, usually small dimensions and always mixed with rain
Ice needles	Poor or moderate	Ice crystals in the form of needles, columns or flakes

Rainwater can be characterized not only in terms of quantity but also qualitatively. Rainwater moving through the atmospheric air layers is contaminated as it deposits significant amounts of dust and aerosols suspended in the air. Then it is further polluted during its runoff from various types of surfaces. Rainwater quality is characterized by high spatial and temporal variability, and its composition depends on many factors, among which the most important are (Zdeb et al. 2018):

- atmospheric pollution,
- catchment type,
- land use, and
- local microclimate.

The most polluted rainwater occurs in urbanized areas, which is mainly associated with the emission of pollutants from the combustion of fossil fuels, industrial plants, and significant traffic. However, it is commonly believed that rainwater coming from housing estates, especially from roofs, is relatively clean. The quality of rainwater decides about the possibilities and purposes of its use. Therefore, these waters, due to the low degree of pollution, that are perceived as a potential alternative source of water.

The quality of collected rainwater from roofs depends both on the type of roofing and its construction (EN 16941-1:2018), as well as on environmental conditions, mainly local climate and atmospheric air pollution (Lee et al. 2010). Potential sources of roof water pollution can be divided into external and internal ones. The external sources include air pollution and organic substances, e.g., leaves and droppings of birds and other animals that may be on the roof. The internal sources of pollution originate from the roofing materials themselves, since there are physicochemical reactions between rainwater and roofing materials (Lee et al. 2012). Numerous studies conducted in the world have shown that the degree of contamination of collected rainwater depends on the type of roofing materials, slope, and length of the roof. Researchers analyzed the quality of rainwater based on microbiological, physical, and chemical parameters.

Rainwater can, in particular, flush heavy metals such as cadmium, copper, lead, zinc, and chromium from roofing materials (Quek and Förster 1993; Melidis et al. 2007; Despins et al. 2009; Mendez et al. 2011; Olaoye and Olaniyan 2012). The research carried out in Texas by Chang and Crowley (1993) for rainwater flowing from 4 different roofing materials showed that as many as 7 metal concentrations exceeded the surface water quality standards set by the United States Environmental Protection Agency. The best quality was characterized by rainwater coming from the terracotta clay roof, and the worst of the wooden shingle roof. Similar comparative studies on 4 different roof materials (concrete, asphalt, ceramic tiles, vegetated roof) were carried out in China (Zhang et al. 2014). Based on them, it was found that the ceramic tile was the most suitable for rainwater harvesting applications. In addition, Zhang et al. (2014) analyzed seasonal trends in water quality parameters, which were shown in winter and spring. In turn, researchers from New Zealand found that in 14% of rainwater samples the permissible concentration of lead was exceeded, which was caused by its leaching from paints covering roofs (Simmons et al. 2001). Other studies also showed that the age and condition of the roof affect negatively the amount of metals leached by rainwater (Chang et al. 2004). Kingett (2003) discovered higher concentrations of zinc in rainwater collected from damaged painted galvanized iron roofs, compared to those in excellent condition. Nicholson et al. (2009) compared harvested rainwater quality among six roof types: galvanized metal, cedar shake, asphalt shingle, two types of wood, and green roof. They noted that the highest concentration of copper and zinc was for the treated woods roofs and the galvanized metal roof.

In addition to contaminants coming directly from roofing materials, rainwater can rinse off any dirt accumulated on the roof surface. The research detected a number of organic compounds in harvested rainwater, including polycyclic aromatic hydrocarbons and pesticides (Basheer et al. 2003; Polkowska et al. 2000; Zobrist et al. 2000). In addition, rainwater is also prone to contamination with microbial aerosols, which may contain about 1800 different types of bacteria (Brodie et al. 2006). In rainwater discharged from the roofs of buildings can also be contaminated with feces of animals such as birds, lizards, and squirrels. Researchers analyzing rainwater samples detected total coliform (TC), fecal coliform (FC), Salmonella spp., Campylobacter, Escherichia coli, Cryptosporidium, and Giardia (Lee et al. 2017; Ahmed et al. 2008). As in the case of physicochemical contamination, changes in seasons (Vialle et al. 2011) are also noticeable in the concentration of microbial contaminants.

The research on the quality of rainwater shows that the first-flush of roof rainwater, which occurs at the beginning of the rainfall, most often contains impurities in increased concentrations. Therefore, separating the first-flush runoff from further rainwater runoff from the roof can result in a significant improvement in harvested water quality (Villarreal and Dixon 2005; Mendez et al. 2011; Gikas and Tsihrintzis 2012).

Therefore, the effective implementation of rainwater harvesting systems requires local research on the physical, chemical, and microbiological properties of rainwater in order to minimize risks to human health and proper selection of water treatment systems (Chong et al. 2013). This applies mainly to cases in which rainwater will be used as potable water and water for bathing and washing.

4.1.2 Technical Aspects of the Use of Rainwater

Rainwater management systems are most often used when rainwater is collected from the roof of the building. This is due to the relatively low level of rainwater pollution compared to other types of terrain. Any rainwater harvesting system can be described through four main functional elements: collection, treatment, storage, and distribution (EN 16941-1:2018). In this case, the basic elements of the installation for the use of rainwater are (Stec and Słyś 2016) the following:

- a roof drainage system (gutters and downpipes),
- devices for rainwater treatment,
- lower storage tank with an overflow system,
- upper storage tank,
- pumping system,
- water supply system supplementing rainwater shortages,
- installation distributing rainwater in the building, and

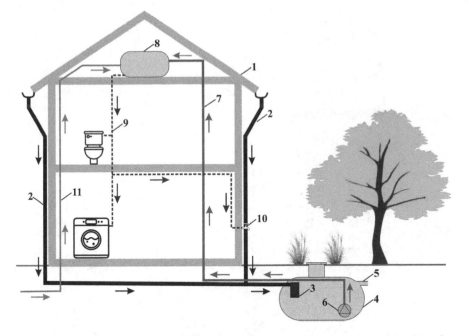

Fig. 4.1 Scheme of rainwater harvesting system, 1—roof panel, 2—gutters and downpipes, 3—pretreatment device, 4—lower storage tank, 5—emergency overflow, 6—pumping system, 7—installation supplying the upper tank, 8—upper storage tank, 9—installation supplying internal sanitary fittings, 10—installation supplying the external water intake point, 11—supplementary installation with water from the water supply network (based on Słyś 2013)

- measuring, regulating, and anti-contamination fittings.

The general scheme of the rainwater harvesting system in a residential building is shown in Fig. 4.1.

Due to individual technical and hydraulic parameters, a rainwater harvesting system can be configured in the following configurations (Słyś 2013):

- configuration 1—installation with a flow tank and discharge of excess water to the sewage system (Fig. 4.2),
- configuration 2—installation with a flow tank and devices for infiltration of excess water (Fig. 4.3),
- configuration 3—installation with a diverter valve and drainage of excess water to the sewage system (Fig. 4.4),
- configuration 4—installation with a diverter valve and devices for infiltration of excess water (Fig. 4.5), and
- configuration 5—installation with a storage tank to take over all rainwater (Fig. 4.5).

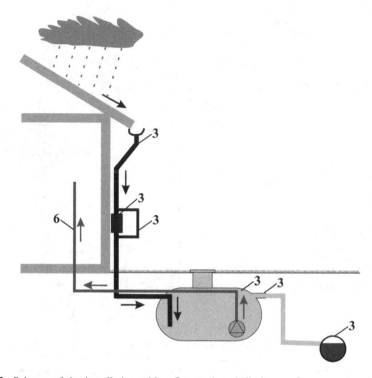

Fig. 4.2 Scheme of the installation with a flow tank and discharge of excess rainwater to the sewer system, 1—gutter and downpipes, 2—filter, 3—filter bypass, 4—storage tank, 5—overflow of excess rainwater, 6—sewage network, 7—water supply to be used (based on Słyś 2013)

In configuration 1 of the installation (Fig. 4.2), rainfall is discharged from the roof with downpipes equipped with filters, whose task is to separate impurities before introducing water into the storage tank. The mechanical cleaning of rainwater is very important as it limits the inflow of organic substances to the tank whose distribution adversely affects the quality of the water held in the tank. The filter can be additionally equipped with a bypass to allow drainage of water from the roof, bypassing the filter, in the situation of its mechanical blockage by contamination. The storage tank holds rainwater, and the excess is directed by the emergency overflow to the sewerage network.

The installation with a flow tank and devices for infiltration of excess rainwater (configuration 2) is a modification of the system shown in Fig. 4.2. The main difference is that the excess of water is discharged from the tank to infiltration facilities. From the point of view of reducing the operating costs of rainwater harvesting, this variant deserves special attention as it is not necessary to make an expensive sewage connection to the network. The tank has a flow-through character, which means that only excess water flows out of it during its full filling, and the remaining water volume is delivered for use. The scheme of such an installation is shown in Fig. 4.3.

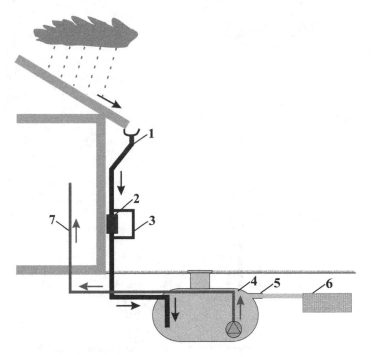

Fig. 4.3 Scheme of an installation with a flow tank and a device for infiltration rainwater to the ground, 1—gutters and downpipes, 2—filter, 3—filter bypass, 4—storage tank, 5—overflow of excess rainwater, 6—drainage device, 7—lead waters to be used (based on Słyś 2013)

In turn in configuration 3 of the installation, the filter has been replaced by a distribution valve whose task is to separate the flow of rainwater into the part discharged directly to the tank and the part directed outside the tank to the sewage system (Fig. 4.4). The valve separates mechanical impurities and discharges them together with a part of rainwater to the sewage system. The efficiency of the distributor operation decreases with the increase of the intensity of flowing rainwater. Therefore, in the course of intense rain, the amount of water supplied to the tank may be lower than the amount of lower intensity applied during the rainfall. This system also does not guarantee satisfying the demand for rainwater. Thus, it is particularly suitable for buildings with a large roof area and a relatively small demand for rainwater.

Similar to the solution shown in Fig. 4.4, the rainwater harvesting system is an installation with a distributor valve for draining excess water directly into the rainwater drainage device (configuration 4). This installation can be used with a significant inflow of rainwater from the roof and a small storage tank capacity, and additionally enables groundwater supply (Fig. 4.5).

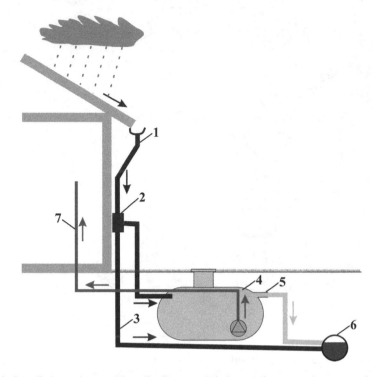

Fig. 4.4 Installation scheme with a distributor and drainage of excess rainwater to the sewage system, 1—gutters and downpipes, 2—distribution valve, 3—drainage of excess water to the sewage system, 4—storage tank, 5—overflow of excess rainwater, 6—sewage network, 7—bringing water for use (based on Słyś 2013)

Configuration 5 of the system is the one where a tank is used to store all rainwater flowing down from the roof of a building and using it (Fig. 4.6). The emergency pipe that the tank is equipped with is very rare, only in incidental cases of tank overflow. Therefore, the best solution is to drain the excess water to the drainage devices. Such a system variant requires the use of a tank with a much larger capacity than in the other installation systems shown.

It should be noted that the choice of rainwater harvesting system configuration has a very significant impact on the economic efficiency of this system. This is due to the design of various tank capacities, which in turn affects capital expenditure, operating costs of the system, fees for sewage disposal, saving tap water and the ability to meet the needs for water of reduced quality, and consequently the economic performance of the system.

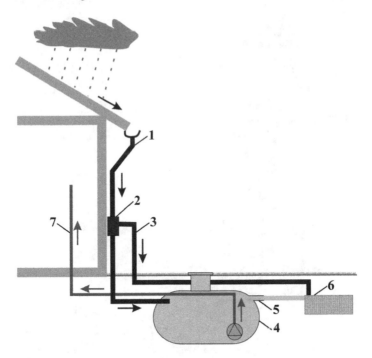

Fig. 4.5 Installation scheme with a distributor and drainage of excess rainwater to the drainage device, 1—gutters and downpipes, 2—distribution valve, 3—excess water drainage to the infiltration device, 4—storage tank, 5—overflow of excess rainwater, 6—drainage device, 7—bringing water for use (based on Słyś 2013)

The basic criterion that determines the layout and equipment of the rainwater commercial installation is the place and scope of use of these waters. In the case of residential buildings, one can distinguish external systems, i.e., those in which rainwater is intended for use outside the building, for instance, for watering greenery, car washing, driveway cleaning, etc. and internal and mixed systems. Rainwater inside the building is mainly used for flushing toilets, laundry, and cleaning work.

Properly designed and constructed installation for the economic use of rainwater will function without problems for many years. That is why it is so important to choose the right elements. The heart of the system is the tank along with the pumping system. Currently, many tank solutions are available in the market that differs in material, construction, strength, and shape. The most common are tanks made of plastic, which can be used as ground or underground tanks. Depending on the purpose of the rainwater system's economic use, storage tanks have suitably adapted equipment, such as a pump, a filter, a siphon, a water intake point, an emergency overflow, and an infusion valve. Figure 4.7 presents an example of a system solution, where rainwater is used outside the building. In the case of difficult mounting conditions, it is a great solution to use flat tanks (Fig. 4.8).

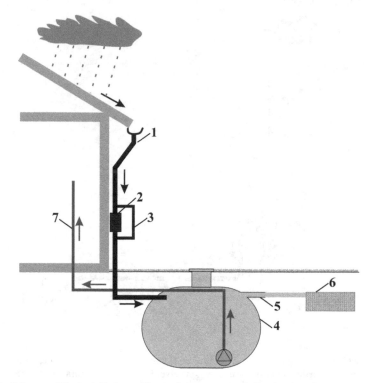

Fig. 4.6 Scheme of the installation with a tank for storage of all rainwater, 1—gutters and down-pipes, 2—filter, 3—filter bypass, 4—storage tank, 5—overflow of excess rainwater, 6—drainage device, 7—water supply to be used (based on Słyś 2013)

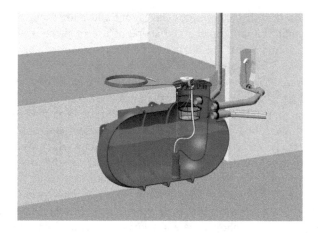

Fig. 4.7 Garden system by MPI company (materials from MPI s.c.)

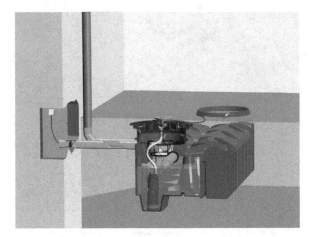

Fig. 4.8 Garden system with a flat tank by MPI company (materials from MPI s.c.)

In the case of internal and mixed systems, the stored rainwater is used in the building mainly for toilets flushing, laundry, and cleaning work. This solution requires a separate water supply system that will supply rainwater to the dredging points. Therefore, the decision to install a stormwater economic system is good at the stage of designing a residential building. Such a system, in addition to all elements of the external system, is equipped with a control system, whose task is to regulate the operation of the installation, especially during rainless periods, when it must be additionally supplied with tap water. An example of a home and garden system solution is shown in Fig. 4.9.

The use of the control panel is a very comfortable solution. The control unit oversaw the solenoid valve responsible for filling the tank with water from the water supply network after reaching the minimum level. The device also controls the pump's operation and protects it against dry running. Through the water-level sensor, the information on reaching the minimum level of filling the tank is transferred to a microprocessor system that activates the water supply solenoid valve. During normal operation of the installation, rainwater is collected from the tank, and if it is

Fig. 4.9 House and garden system with a control center (materials from GRAF company)

Fig. 4.10 Rainwater harvesting system with central switching station for direct water intake from the water supply network (materials from MPI s.c.)

not installed, the solenoid valve completes the tank with drinking water until the maximum state is reached.

The system of economic use of rainwater can also be equipped with a rainwater unit enabling automatic switching to direct consumption of tap water in periods of rainwater shortage. The general scheme for such an installation is shown in Fig. 4.10. The control panel monitors the system in such a way that it is ready for commissioning at any time. The operation of the three-way valve is controlled by means of a float switch located in the rainwater tank. The control panel protects the pump against dry running and allows it to be switched off when a certain pressure in the system is reached.

In addition to systems with underground external tanks, systems with a tank located inside the building, for example in the basement (Fig. 4.11) or outside in the garden (Fig. 4.12), can also be used. In the latter case, rainwater is used only for watering greenery.

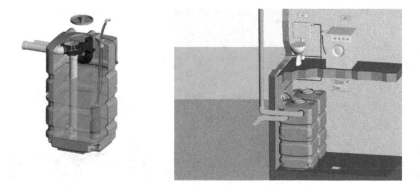

Fig. 4.11 Rainwater harvesting system with an internal tank (materials from MPI s.c.)

Fig. 4.12 Garden system for
collecting rainwater
(materials from MPI s.c.)

According to the European Standard EN 16941-1:2018, the treatment system shall

- be water resistant and durable,
- be accessible for maintenance,
- not affect the hydraulic operation of the overall drainage system, and
- have a hydraulic efficiency ratio of at least 0.9.

A very important element of the rainwater system is the filtration process, which can be implemented on filters mounted on downpipes, filters installed in the ground or in the tank itself. The first group is distinguished by rainwater collectors with a filter or a filter installed inside and a valve. Devices of this type are mounted on downpipes in order to stop thicker impurities flowing from the roof, mainly leaves.

For the separation of larger pollutants from the rainwater, gutter leaf separators are also used. This device separates the leaves and larger impurities from the water and then removes them from the separator. Thanks to the leaf separator, the drain pipes will not block as the dirt accumulating in them is filtered on the device continuously.

A separate group of devices used to remove contaminants from rainwater is underground filters installed in the ground in front of a rainwater storage tank. The filter basket prevents the contamination particles from entering along with the rainwater into the tank. Very small filtration holes of the basket ensure a high degree of water purification, so that they can be used even in home systems.

The task of tank filters, as well as external filters, is to remove impurities (leaves, dust, pollen from plants) that have been rinsed from the roof. In the case of greater rainwater pollution, it is possible to use a two-stage filtration system. The rainwater brought to the reservoir flows into the first of the mesh filter cartridges, from which large impurities are discharged to the sewage system, whereas the pretreated water is directed to the second filter cartridge with smaller holes. On it, the water is further purified and the contaminated pollutants are transported to the sewage system, and the filtered water flows into the tank.

In home rainwater systems, filters are increasingly used to trap impurities in suspended water. The filter construction forces the centrifugal vortex movement of water, thanks to which impurities fall to the bottom of the diffuser, while the filtering sleeve ensures final filtration in accordance with the selected purification accuracy. The filter cover from its bottom is equipped with a drain plug designed to periodically remove larger impurities. It is a filter that is mounted on the water inlet to the installation in the building. This location guarantees the protection of all devices connected to the water system. The extensive version of this filter purifies water in two stages. In the first one, water flows through a mechanical filter where larger solid contaminants suspended in water (particles of soil, sand, sediments, etc.) are retained, while in the second through a filter with active carbon or polyphosphate. Depending on the selected deposit, this filtration step can soften water or improve its organoleptic qualities.

In order to broaden the range of rainwater applications and to increase the safety of its use, filtering systems are used in combination with ultraviolet radiation. It is an ideal method of removing impurities from water, among others such as bacteria, viruses, and protozoa. Rainwater of this quality can be safely used for washing.

4.2 Graywater Recycling

According to the European Standard PN-EN 12056-1, gray water is wastewater free of urine and feces. They are created every day during washing the body, washing dishes, and washing. Distribution of water used for gray water and black sewage, which come from the toilet bowls flushing, enables the reuse of the former. In addition, gray water can be divided into so-called gray and light gray water (Gross et al. 2015). Light gray water usually does not include wastewater from dishwashing. Some also do not include the gray water from washing.

In both cases, the used water is treated as dark gray water. Reuse of gray water allows for at least a twofold use of water, and in some cases even three times, which results in savings in the range of 10–20% of total water consumption in the city (Gross et al. 2015).

4.2.1 The Characteristics of Gray water

Generally, impurities in wastewater can be divided into three groups: physical, chemical, and biological. The first group is mainly a suspension in the form of solid, insoluble, and organic and inorganic solids. Chemical impurities include proteins, fats, and carbohydrates, as well as mineral substances. Biological pollutants, in turn, are determined by the content of viruses, bacteria, fungi, and parasitic eggs in the wastewater.

An important physical parameter of gray water is their temperature. It depends on the source of their formation and varies between 18 and 38 °C (Eriksson et al. 2002). Higher temperature results from using warm and hot water for bathing. Under favorable conditions, higher temperature of gray water is undesirable as it may intensify the development of microorganisms.

Gray water differs significantly from black sewage both in quantity and in diversity of contaminants contained in them (Marleni et al. 2015). These wastewater contain quickly decomposable organic substances (Jefferson et al. 2004). This is confirmed by the study quoted by Iwanicka (2012), which showed that biochemical oxygen demand (BOD_5) after 5 days decreases by almost 90%, and in black wastes by only 40% (Iwanicka 2012).

The concentration of individual pollutants for gray water is conditioned mainly by the source of their formation (Vakil et al. 2014), as well as the lifestyle, social and cultural behavior of residents, and the availability of water and its consumption (Ghaitidak and Kunwar 2013). It can vary widely (Donner et al. 2010), however, wastewater coming from the kitchen is considered the most contaminated. They often contain oils and other unfavorable substances whose presence in gray water disqualifies their use for toilet bowls flushing or watering the garden (Oron et al. 2014). These wastewaters are characterized by the highest oxygen demand from all of their sources. As research from different countries shows, the concentration of BOD_5 may vary within a wide range, e.g., from 5 to 1460 mg/dm^3 (Eriksson et al. 2002), 1850 mg/dm^3 (Abu-Ghunmi et al. 2008) or from 470 up to 4450 mg/dm^3 (Ghaitidak and Kunwar 2013). Kitchen wastewater can also contain various types of microorganisms derived from washing vegetables and fruits or raw meat. The concentration of E. coli in such wastewater was observed as 1.3×10^5 and 2.5×10^8/100 ml, and fecal streptococci between 5.15×10^3 and 5.5×10^8/100 ml (Eriksson et al. 2002). Some authors (Li et al. 2009), however, recommend that wastewater discharged from sinks and dishwashers should be mixed with other types of gray water in case they are to be subjected to biological treatment.

Sewage generated during washing clothes, in turn, contains significant amounts of detergents and bleaches. Their quality changes during the washing cycle and depends on the type of detergents used. This type of gray water is characterized by significant salinity and the content of fibrous suspension derived from clothes. Although research is being conducted on their use (Misra and Sivongxay 2009; Misra et al. 2010), they still raise a lot of controversy, especially that gastrointestinal pathogens may also be present in these wastewater (O'Toole et al. 2012). Bacteria in wastewater from washing machines occur mainly in the case of washing clothes of small children or sick people. The total number of fecal Coli in these wastewaters can be at the level of 1.6×10^4 and 9×10^4/100 ml and fecal streptococci in the range 1×10^6 and 1.3×10^6/100 ml (Eriksson et al. 2002).

Graywater from bathrooms contains soap, shampoos, cleaning products, and hair, and in the case where children or the elderly are present in the household, also traces of phlegm substances (Ottoson and Stenström 2003). However, the total concentration of pollutants is relatively small in comparison to sewage discharged from other sanitary facilities, which allows their reuse after the basic pretreatment process. Pollutants contained in them are easily decomposable, so the time of their retention should not be long, as they may rot. Sewage generated from the bath is characterized by a BOD$_5$ concentration of 50–300 mg/dm^3 and a suspension content of 43–120 mg/dm^3 (Li et al. 2009). Bathroom gray sewage contains up to 3 × 10^3/100 ml of fecal coliforms and fecal streptococci in the range of 1–7 × 10^4/100 ml (Eriksson et al. 2002).

Gray water generated during bathing and hand washing is the most commonly used in Europe. In Germany and the Netherlands, sewage from laundering is also often used (Iwanicka 2012). Mixed gray water is characterized by lower fluctuations in pollution concentrations in relation to wastewater from a single source. This is due to the leveling of pollutants of various types of gray water during their storage in one tank. Such sewage is characterized by a suspension content of 6.4–240 mg/dm^3, a biochemical oxygen demand in the range 50–350 mg/dm^3, and a fecal bacterial concentration from 1 to 1.5 × 10^8 (Li et al. 2009).

The amount of gray water generated in a given building depends only on the water consumption by residents for a given purpose. As shown in Chap. 3, the demand for water in residential buildings varies widely and from country to country. When analyzing the ratio of quantity of graywater flow (Q_g) to water consumption (Q_w), it can be seen that its value is also significantly changed. Ghaitidak and Kunwar (2013) reported that it could range from 0.31 to even 0.87 in low-income countries. The number of gray water produced in a given source is also determined by the lifestyle and inhabitants' welfare. It is generally accepted that gray water discharged from body washing constitutes to 47%, produced in the kitchen 27%, and from washing clothes 26% (Ghaitidak and Kunwar 2013).

4.2.2 Graywater Recycling Systems

In the vast majority of cases graywater is used for nonfood purposes, not having direct contact with people, such as toilets flushing, watering the garden, and washing the car. However, utility, hygienic, and sanitary safety reasons make it necessary to clean them. Knowledge of the type and concentrations of contaminants is required when selecting the appropriate treatment technology that allows the reuse of gray water in buildings.

The reduction of turbidity and the content of suspended solids in gray water allows preventing deposits in the devices in which they are used. On the other hand, a reduction in the concentration of substances undergoing oxidative decomposition is necessary to minimize the risk of their killing. Even in situations where there is no direct contact between gray wastewater and people, it is also important to remove bacteria, because, in the course of using sanitary facilities, sewage may splash and then infect people through airways or open wounds (Gross et al. 2007; Benami et al. 2016).

Depending on the place where gray water is used, more or less advanced recycling systems are applied to enable the removal of pollutants, storage of pretreated sewage and their installation. Many different graywater recycling systems have been implemented all over the world. They use both simple pollution removal technologies (filters, biofilters) as well as advanced membrane technologies (Gross et al. 2015).

There are no uniform international guidelines defining the quality and scope of gray wastewater utilization. In many cases, national and sometimes even local recommendations regarding the use of graywater recycling systems are established. They focus mainly on the health and the impact of these systems on the natural environment. These guidelines define primarily the permissible concentrations of pollutants in treated gray sewage, the value of which depends on the place where sewage is used. Table 4.1 shows examples of standards from different countries.

For European Union Member States, the European Council Directive 91/271/EEC on urban wastewater treatment applies. It states that "treated wastewater shall be reused when appropriate," but it does not explain exactly what it means and to what extent treated wastewater can be reused. Detailed guidelines for the graywater systems were developed in Great Britain in 2010 (BS 2010). They define the possibilities of using graywater, technical solutions, and threats resulting from their use. Germany is the leader in the use of gray water. Domestic graywater reuse systems are legal in Germany, but must be registered with the Health Office to ensure no connection to the potable water network and ensure correct labeling according to this regulation (Allen et al. 2010). The United States does not have a general national policy, leaving the gray water to the state. About 30 of the 50 states have graywater regulations (Allen et al. 2010). The situation is similar in Australia (CEM 2013).

Taking into account the graywater recycling systems, there are several basic types (BS 2010):

- direct reuse system,
- short retention system,
- basic physical and chemical system,
- biological system,
- biomechanical system, and
- hybrid system.

Direct reuse system is a solution where there are no graywater treatment processes and they are not retained, or the time of their storage is very short. This is dictated by the rapidly deteriorating quality of untreated gray water. They are simple devices to collect the gray water and direct them to the points of use. In such systems, the use of gray water is limited only to subsurface irrigation and non-spray applications (BS 2010) (Table 4.2).

Another type of graywater recycling system is a system with short wastewater retention. The system uses very basic cleaning techniques such as filtration and scraping of dirt from the surface of gray water and allowing the deposition of particles on the bottom of the tank. Their goal is mainly to reduce the problems associated with unpleasant odor and to improve the quality of the gray water, to the extent that will allow them to briefly survive.

Basic physical and chemical system is a system where wastewater filtration processes are used before being introduced into the retention reservoir. Then chemical disinfection takes place in the tank, limiting the growth of bacteria during the storage of the gray water.

Table 4.2 Graywater reuse standards from different countries (based on Li et al. 2009; CEM 2013; BS 2010)

Country	pH	BOD$_5$ [mg/dm^3]	Turbidity [NTU]	Fecal coliforms	Total coliforms	Reuse application
China	6–9	<10	<5	<3/100 ml	–	Toilet flushing
	6–9	<20	<20	<3/100 ml	–	Irrigation
	6–9	<6	<5	<3/100 ml	–	Washing
Germany	6–9	5	1–2	<10 ml	<100 ml	Toilet flushing
Japan	5.8–8.6	≤20	–	–	<1000 ml	Toilet flushing
	5.8–8.6	≤20	–	–	≤50 ml	Irrigation
The USA	6–9	≤10	<2	Non-detectable/100 ml	–	Toilet flushing Irrigation Car washing
The UK	5–9.5	–	<10	<250/100 ml	<1000 ml	Toilet flushing
	5–9.5	–	–	<250/100 ml	<1000 ml	Irrigation
	5–9.5	–	<10	Non-detectable/100 ml	<10 ml	Washing
Singapore	6–9	<5	<2	Non-detectable/100 ml	<10/100 ml	Toilet flushing Irrigation Washing

In biological systems aerobic or anaerobic bacteria are used, whose task is to degrade organic matter contained in gray water. In the case of aerobic treatment, pumps or aquatic plants can be used to aerate the water.

Biomechanical systems are the most advanced solutions for reusing gray water at home. They combine the biological and physical precessions of wastewater treatment, e.g., the removal of organic matter by bacteria and solid substances by sedimentation.

Figure 4.13 shows a typical wastewater recycling system with indirect water supply from the water supply network. In this solution, there is an upper tank to which cleaned gray water and water from the water supply network are supplied. The process of filling the tank is controlled by a control and measurement system. Gray water coming from the body wash comes to a purification unit equipped with a pump forcing pretreated sewage into the upper tank, from where it is fed to non-potable applications, e.g., WC, washing machine, and garden tap.

In turn, Fig. 4.14 shows the system layout with the use of gray water and direct water supply from the water supply network to the control module. This module is equipped with a pump that takes purified gray sewage from the purification unit and then pushes it to non-potable applications. A similar solution with direct drinking water supply is shown in Fig. 4.15. However, in this system there is only one device, which is both a purifying unit and a pumping and control system.

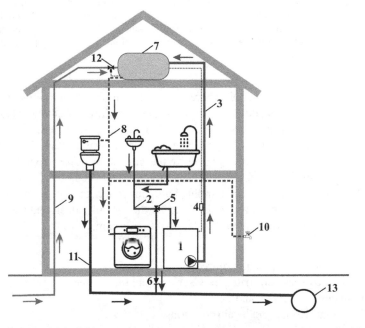

Fig. 4.13 System with indirect treated graywater supply, 1—graywater treatment unit with a pump, 2—pipes carrying bathroom gray water for treatment, 3—pipes carrying treated gray water to upper tank, 4—control panel, 5—bypass, 6—anti-surcharge valve, 7—upper tank, 8—distribution pipes carrying treated gray water, 9—drinking water supply pipe, 10—garden tap, 11—pipes discharging blackwater to the sewage system, 12—electronic control, 13—sewage system (based on BS 2010)

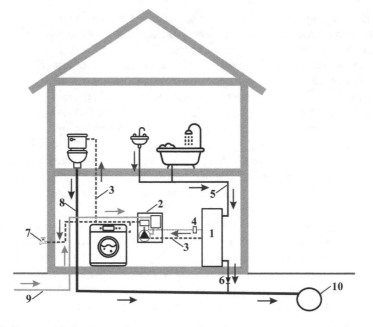

Fig. 4.14 System with direct treated graywater supply and a separate treatment unit, 1—graywater treatment unit, 2—module with a pump, 3—pipes carrying treated gray water, 4—control panel, 5—pipes carrying bathroom gray water for treatment, 6—anti-surcharge valve, 7—garden tap, 8—pipes discharging blackwater to the sewage system, 9—drinking water supply pipe, 10—sewage system (based on BS 2010)

Hybrid systems are also used, where there are various combinations of the solutions described above. In addition, the graywater recycling system can also be integrated into the rainwater harvesting system. Thanks to this, in the rainless periods or with little rainfall it is possible to further reduce the water intake from the water supply network, which will be replaced by the gray water. The integrated solution layout is shown in Figs. 4.16 and 4.17. Integrated systems can be operated as separate, independent installations or combined into one power source. Such installations are considered to be specialized and should be thoughtfully designed and correctly made. The following issues should be taken into account:

- variability of tributaries,
- type of cleaning equipment required,
- managing the excess of treated gray sewage and rain water,
- environmental concerns, and
- applicable guidelines and legal provisions.

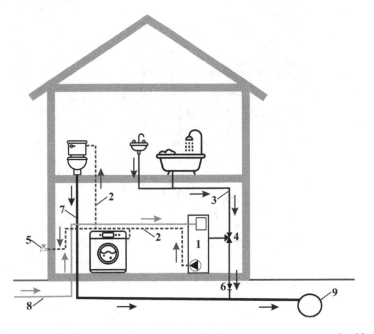

Fig. 4.15 System with direct treated graywater supply, 1—graywater treatment unit with a pump, 2—pipes carrying treated gray water, 3—pipes carrying bathroom gray water for treatment, 4—bypass, 5—garden tap, 6—anti-surcharge valve, 7—pipes discharging blackwater to the sewage system, 8—drinking water supply pipe, 9—sewage system (based on BS 2010)

The technologies used for graywater treatment operate on physical, chemical, and biological processes. Most of them are preceded by a preliminary stage involving the separation of solids in septic tanks, filter bags, and various types of sieves and filters. Thanks to this, the clogging phenomenon is limited in the proper purification phase. The final stage is the disinfection of gray sewage to meet the microbiological requirements (Li et al. 2009).

The physical purification includes coarse sand and soil filtration and membrane filtration. Physical processes have little effect on improving the quality of gray water, therefore it is not recommended to use them only in recycling systems (Li et al. 2009). Chemical processes can effectively remove suspended solids, organic substances, and surfactants from gray wastewater. Among them, it is possible to use. coagulation, photo-catalytic oxidation, granular activated carbon, and ion exchange. In turn, the effectiveness of biological processes depends on the conditions in which they occur. Due to the low reduction of both organic substances and surfactants, anaerobic processes are not recommended for the purification of the gray water (Li et al. 2009). Biological oxygen processes that are most often used in gray wastewater recycling systems are sequencing batch reactor (SBR), rotating biological contractor (RBC), and constructed wetland. In order to meet the standards specified for treated gray water, most of the biological processes still require a filtration and/or disinfection step. The exception is the membrane bioreactor (MBR), in which biological processes

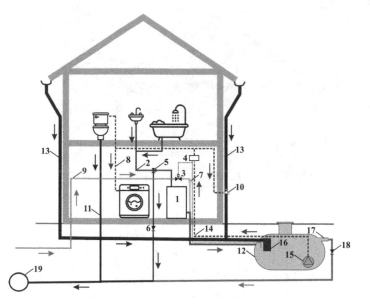

Fig. 4.16 Integrated gray water and rainwater system with direct non-potable supply, 1—graywater treatment, 2—pipes carrying bathroom graywater for treatment, 3—electronic control, 4—control panel, 5—bypass, 6—anti-surcharge valve, 7—air gap, 8—distribution pipes carrying treated grey-water/rainwater, 9—potable water supply pipe, 10—garden tap, 11—pipes discharging blackwater to the sewage system, 13—gutter, 14—rainwater tank, 15—submersible pump, 16—filter, 17—overflow to sewage system, 18—anti-surcharge valve and vermin trap, 19—sewage system (based on BS 2010)

and membrane filtration take place. MBR is the only technology that allows satisfactory removal of organic substances, surfactants, and microbiological contaminants without a filtration and disinfection step.

Considering the above, combining the aerobic biological process with physical filtration and disinfection is considered the most optimal solution for recycling gray water, not only in terms of the level of pollution reduction but also economic (Li et al. 2009).

Li et al. (2009) proposed different options for dealing with gray water depending on their quality and place of use. The diagram illustrating these procedures is shown in Fig. 4.18. It shows that regardless of the type of gray water the first stage is their pretreatment by sedimentation of pollutants and separation of substances lighter than water. The subsequent stages should be adapted to the requirements for treated gray water. In the case of low strength, the gray water is a chemical, physical cleaning on filters, and possibly disinfection. However, if they are medium and high strength gray water, biological methods of removing impurities, filtration, and in some situations chlorine or UV-disinfection are used. These processes make it possible to meet the requirements with regard to the contents in the sewage of purified microorganisms, suspended solids, and turbidity. Gray water after its purification in membrane bioreactor (MBR) is characterized by a low degree of pollution and meets

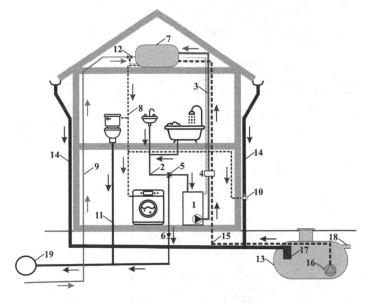

Fig. 4.17 Integrated gray water and rainwater system with indirect non-potable supply, 1—gray water treatment unit with a pump, 2—pipes carrying bathroom gray water for treatment, 3—pipes carrying treated gray water to upper tank, 4—control panel, 5—bypass, 6—anti-surcharge valve, 7—upper tank, 8—distribution pipes carrying treated gray water/rainwater, 9—potable water supply pipe, 10—garden tap, 11—pipes discharging blackwater to the sewage system, 12—electronic control, 13—rainwater tank, 14—gutter, 15—pipes carrying treated rainwater to upper tank, 16—submersible pump, 17—filter, 18—overflow to rainwater management system, 19—sewage system (based on BS 2010)

the most stringent standards thanks to which virtually unlimited use is possible for non-potable urban reuses (Li et al. 2009).

Various gray waste recycling systems are available in the market, whose design solutions and functioning are based on the purification methods shown in Fig. 4.18. These systems are adapted to the required efficiency by selecting the appropriate capacity of tanks resulting from the demand for non-potable water. An example of such a system is shown in Fig. 4.19. It is a system in which biological purification and ultrafiltration on a membrane filter (MBR bioreactor) has been applied. After treatment, the water flows to the storage tank from where it is pumped to the installation. This system guarantees complete separation of the biomass, thanks to which the water is clear and free from bacteria and viruses. The system operation is controlled by an electronic control panel. A similar solution, but with much greater efficiency is shown in Fig. 4.20. The required system capacity was achieved by combining several tanks. In both cases, purified gray water can be used for flushing toilets, irrigation, cleaning, and washing.

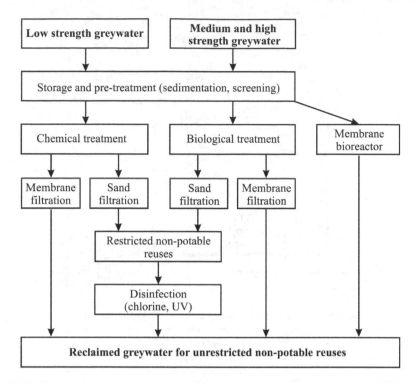

Fig. 4.18 The graywater recycling schemes for non-potable uses (based on Li et al. 2009)

Figure 4.21 presents a solution in which the gray water is biologically purified and then disinfected with UV radiation. This system works fully automatically. The cleaned gray water does not pose hygienic risks and is suitable for toilets flushing and watering gardens.

The success of using graywater recycling systems depends on many factors, including technical, economic, and social ones. Due to the variability in the quantity and quality of gray water, there are no universal systems suitable for each building. Therefore, the implementation of these systems should be adapted each time to local conditions. The key factor for the safe use of pretreated gray water is compliance with legal provisions specifying the possibilities of wastewater recycling in a given country, as well as compliance with the rules to minimize risks and threats to human health.

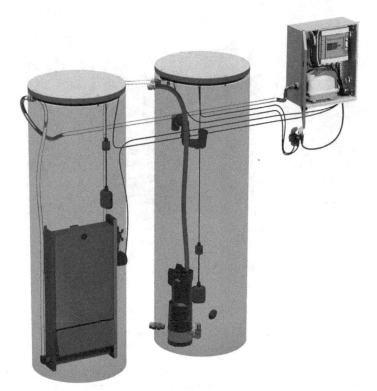

Fig. 4.19 Graywater recycling system with a membrane filter and treatment capacity of up to 250 L/day (GreenLife)

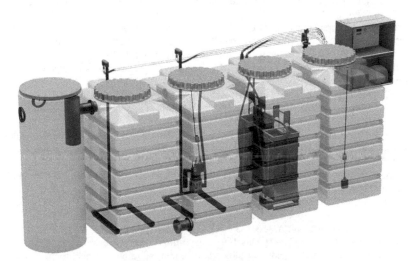

Fig. 4.20 Graywater recycling system with a membrane filter and treatment capacity of up to 2000 L/day (GreenLife)

Fig. 4.21 Graywater recycling system with fixed-bed reactor and UV-disinfection and treatment capacity of up to 750 L/day (GreenLife)

References

Abu-Ghunmi L, Zeeman G, Lier JV, Fayyed M (2008) Quantitative and qualitative characteristics of grey water for reuse requirements and treatment alternatives: the case of Jordan. Water Sci Technol 1385–1396

Ahmed W, Huygens F, Goonetilleke A, Gardner T (2008) Real-time PCR detection of pathogenic microorganisms in roof-harvested rainwater in Southeast Queensland. Aust Appl Environ Microbiol 74(17):5490–5496

Allen L, Christian-Smith J, Palaniappan M (2010) Overview of greywater reuse: The potential of grewater systems to aid sustainable water management. Pacific Institute, Oakland, California. ISBN 1-893790-27-4

Basheer C, Balasubramanian R, Lee HK (2003) Determination of organic micropollutants in rainwater using hollow fiber membrane/liquid-phase microextraction combined with gas chromatography-mass spectrometry. J Chromatogr A 1016(1):11–20

Benami M, Busgang A, Gillor O, Gross A (2016) Quantification and risks associated with bacterial aerosols near domestic greywater-treatment systems. Sci Total Environ 562:344–352

Brodie EL, DeSantis TZ, Moberg Parker JP, Zubietta IX, Piceno YM, Andersen GL (2006) Urban aerosols harbor diverse and dynamic bacterial populations. Proc Natl Acad Sci 104(1):299–304

BS 8525-1:2010, Grey water systems, Part 1: Code of Practice

CEM (2013) Greywater for UK housing. The College of Estate Management, United Kingdom

Chang M, Crowley CM (1993) Preliminary observations on water quality of storm runoff from four selected residential roofs. Water Resour Bull 29(5):777–783

Chang M, McBroom MW, Beasley SR (2004) Roofing as a source of nonpoint water pollution. J Environ Manag 73(4):307–315

Chong MN, Sidhu J, Aryal R, Tang J, Gernjak W, Escher B, Toze S (2013) Urban stormwater harvesting and reuse: a probe into the chemical, toxicology and microbiological contaminants in water quality. Environ Monit Assess 185:6645–6652

Council Directive 91/271/EEC of 21 May 1991 concerning urban waste-water treatment

Despins C, Farahbakhsh K, Leidl C (2009) Assessment of rainwater quality from rainwater harvesting systems in Ontario, Canada. J Water Supply: Res Technol AQUA 58(2):117–134

Donner E, Eriksson E, Revitt DM, Scholes L, Holten Lützhøft HC, Ledin A (2010) Presence and fate of priority substances in domestic greywater treatment and reuse systems. Sci Total Environ 408:2444–2451

EN 16941-1:2018, On-site non-potable water systems—Part 1: systems for the use of rainwater. European Committee for Standardization, Brussels

Eriksson E, Auffarth K, Henze M, Ledin A (2002) Characteristics of grey wastewater. Urban Water 4:85–104

Ghaitidak DM, Kunwar DY (2013) Characteristics and treatment of greywater—a review. Environ Sci Pollut Res 20:2795–2809

Gikas G, Tsihrintzis V (2012) Assessment of water quality of first-flush roof runoff and harvested rainwater. J Hydrol 466–467:115–126

Gross A, Shmueli O, Ronen Z, Raveh E (2007) Recycled vertical flow constructed wetland (RVFCW)—a novel method of recycling greywater for irrigation in small communities and households. Chemosphere 66:916–923

Gross A, Maimon A, Alfiya Y, Friedler E (2015) Greywater reuse. CRC Press

Iwanicka Z (2012) Charakterystyka ścieków szarych. Interdyscyplinarne zagadnienia w inżynierii i ochronie środowiska: praca zbiorowa. Oficyna Wydawnicza Politechniki Wrocławskiej. ISBN 978-83-7493-671-2

Jefferson B, Palmer A, Jeffery P, Stuez R, Judd S (2004) Grey water characterisation and its impact on the selection. Water Sci Technol 50(2):157–164

Kingett M (2003) A study of roof runoff quality in auckland New Zealand: implications for storm water management. Auckland Regional Council, Auckland, New Zealand

Lee JY, Yang JS, Han M, Choi J (2010) Comparison of the microbiological and chemical characterization of harvested rainwater and reservoir water as alternative water resources. Sci Total Environ 408(4):896–905

Lee JY, Bak G, Han M (2012) Quality of roof-harvested rainwater—comparison of different roofing materials. Environ Pollut 162:422–429

Lee M, Kim M, Kim Y, Han M (2017) Consideration of rainwater quality parameters for drinking purposes: a case study in rural Vietnam. J Environ Manag 200:400–406

Li F, Wichmann K, Otterpohl R (2009) Evaluation of appropriate technologies for grey water treatments and reuses. Water Sci Technol 249–260

Marleni N, Gray S, Sharma A, Burn S, Muttil N (2015) Impact of water management practice scenarios on wastewater flow and contaminant concentration. J Environ Manag 151:461–471

Melidis P, Akratos CS, Tsihrintzis VA, Trikilidou E (2007) Characterization of rain and roof drainage water quality in Xanthi. Greece Environ Monit Assess 127:15–27

Mendez CB, Klenzendorf JB, Afshar BR, Simmons MT, Barrett ME, Kinney KA, Kirisits M (2011) The effect of roofing material on the quality of harvested rainwater. Water Res 45:2049–2059

Misra RK, Sivongxay A (2009) Reuse of laundry greywater as affected by its interaction with saturated soil. J Hydrol 366:55–61

Misra RK, Patel JH, Baxi VR (2010) Reuse potential of laundry greywater for irrigation based on growth, water and nutrient use of tomato. J Hydrol 386:95–102

Nicholson N, Clark SE, Long BV, Spicher J. Steele KA (2009) Rainwater harvesting for non-potable use in gardens: a comparison of runoff water quality from green vs. traditional roofs. In: Proceedings of world environmental and water resources congress 2009—Great Rivers Kansas City, Missouri

O'Toole J, Sinclair M, Malawaraarachchi M, Hamilton A, Barker SF, Leder K (2012) Microbial quality assessment of household greywater. Water Res 46:4301–4313

Olaoye RA, Olaniyan OS (2012) Quality of rainwater from different roof material. Int J Eng Technol. ISSN:2049-3444

Oron G, Adel M, Agmon V, Friedler E, Halperin R, Leshem E, Weinberg D (2014) Greywater use in Israel and worldwide: standards and prospects. Water Res 58:92–101

Ottoson J, Stenström TA (2003) Faecal contamination of greywater and associated microbial risks. Water Res 37:645–655

Polkowska Z, Kot A, Wiergowski M, Wolska L, Wolowska K, Namiesnik J (2000) Organic pollutants in precipitation: determination of pesticides and polycyclic aromatic hydrocarbons in Gdansk. Poland Atmos Environ 34(8):1233–1245

Quek U, Förster J (1993) Trace metals in roof runoff. Water Air Soil Pollut 68(3–4):373–389

Simmons G, Hope V, Lewis G, Whitmore J, Gao W (2001) Contamination of potable roof-collected rainwater in Auckland. New Zealand Water Res 35(6):1518–1524

Słyś D (2013) Zrównoważone systemy odwodnienia miast. Dolnosląskie Wydawnictwo Edukacyjne, Wrocław

Stec A, Słyś D (2016) Instalacje ekologiczne w budownictwie mieszkaniowym. Wydawnictwo i Handel Książkami „KaBe" (Ecological installations in residential buildings. „KaBe" Publishing House and Book Trade), Krosno

Sumner G (1988) Precipitation. Process and analysis. John Wiley and Sons, London. ISBN-13: 978-0471905349

Vakil KA, Sharma MK, Bhatia A, Kazmi AA, Sarkar S (2014) Characterization of greywater in an Indian middle-class household and investigation of physicochemical treatment using electrocoagulation. Sep Purif Technol 130:160–166

Vialle C, Sablayrolles C, Lovera M, Jacob S, Huau M, Montrejaud-Vignoles M (2011) Monitoring of water quality from roof runoff: Interpretation using multivariate analysis. Water Res 45:3765–3775

Villarreal E, Dixon A (2005) Analysis of a rainwater collection system for domestic water supply in Ringdansen, Norrkoping. Sweden Build Environ 40:1174–1184

Zdeb M, Papciak D, Zamorska J (2018) An assessment of the quality and use of rainwater as the basis for sustainable water management in suburban areas. E3S Web of conferences 45:00111 https://doi.org/10.1051/e3sconf/20184500111. INFRAEKO 2018

Zhang Q, Wang X, Hou P, Wan W, Li R, Ren Y, Ouyang Z (2014) Quality and seasonal variation of rainwater harvested from concrete, asphalt, ceramic tile and green roofs in Chongqing, China. J Environ Manag 132:178–187

Zobrist J, Müller S, Ammann A, Bucheli T, Mottier V, Ochs M, Schoenenberger R, Eugster J, Boller M (2000) Quality of roof runoff for groundwater infiltration. Water Res 34(5):1455–1462

Chapter 5
Research on the Effectiveness of Systems with Alternative Water Sources for Buildings Located in Selected European Countries

Abstract Research on the efficiency of using the rainwater harvesting system was carried out at eight different locations in Europe. The simulation model used for the research is based on the daily mass water balance. The model algorithm is based on the YAS (Yield After Spillage) operating rule. In the research, it was assumed that rainwater would be used only for non-potable uses. It was assumed that the harvested rainwater would be used for toilet flushing, washing, and garden watering. In order to determine the optimal tank capacity, the volumetric reliability of the series of rainwater tanks which were assumed in the first step was determined. The tank capacity was considered optimal when a further increase in this capacity resulted in changes in volumetric reliability of 1% or less.

5.1 Simulation Model of the Rainwater Harvesting System

Many methods of analysis are available including behavioral models that allow one to determine the required capacity of a rainwater storage tank (Mitchell 2007). There are two main behavioral models describing how rainwater harvesting systems (RWHS) works: yield-after-spillage (YAS) and yield-before-spillage (YBS) (Fewkes 2000).

The YAS operating rule determines the yield as the storaged volume at the end of the previous time interval or the water demand in the current time interval, whichever is smaller. The rainwater runoff is then added to the storage volume of rainwater from the preceding time step and any rainwater excess spilling by the overflow and then subtracts the yield. The yield-before-spillage (YBS) operating rule assigns the yield as the storaged volume of rainwater from the previous time step plus the volume of runoff in the current time or the demand, whichever is smaller (Fewkes 2000). Figure 5.1 shows the difference between how the rainwater harvesting system works in accordance with the two YAS and YBS models.

The research used the simulation model by Słyś (2009), whose calculation algorithm is based on YAS operating rule (5.1) and (5.2), which is the same as recommended in the standard EN 16941-1:2018. Figure 5.2 shows the rainwater harvesting system within the behavior analysis model, which was used in the study.

© Springer Nature Switzerland AG 2020
A. Stec, *Sustainable Water Management in Buildings*,
Water Science and Technology Library 90,
https://doi.org/10.1007/978-3-030-35959-1_5

Fig. 5.1 Functioning of
rainwater harvesting system
according to the YAS and
YBS operational rule (based
on Mitchell 2007)

Fig. 5.2 Rainwater
harvesting system
configuration, R_t—rainfall,
I_t—rainwater runoff,
D_t—rainwater demand,
M_t—mains supply makeup,
Y_t—yield from store,
V_t—rainwater tank storage
volume, C—store capacity,
O_t—rainwater overflow

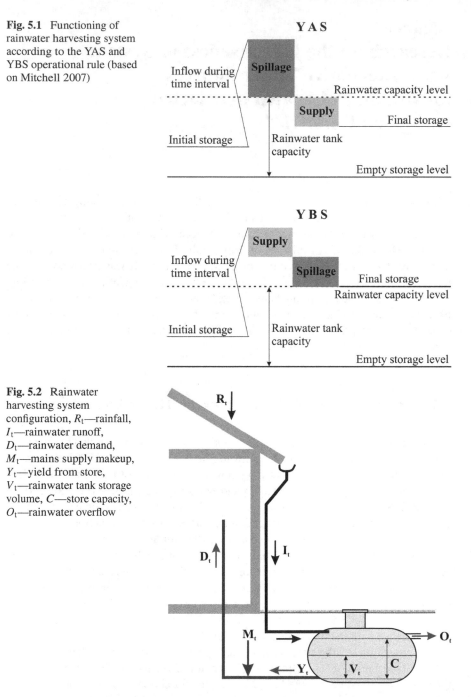

$$Y_t = \min\begin{cases} D_t \\ V_{t-1} \end{cases} \tag{5.1}$$

$$V_t = \min\begin{cases} V_{t-1} + I_t - Y_t \\ C - Y_t \end{cases} \tag{5.2}$$

where Y_t is the yield from store during time interval t, m^3, D_t is the water demand during time interval t, m^3, V_{t-1} is the storage volume at the end of the previous time step, m^3, V_t is the rainwater tank storage volume during time interval t, m^3, I_t is the rainwater runoff during time interval, m^3, C is the store capacity, m^3.

Behavior analysis uses continuous simulation to study changes over time in the volume of the water stored. These changes are calculated on the basis of a mass balance Eq. (5.3) (Mitchell 2007).

$$V_t = V_{t-1} + I_t + R_t - E_t - O_t - L_t - Y_t \tag{5.3}$$

where R_t is rainfall during time interval t, m, E_t is the evaporation from the rainwater tank store during time interval t, O_t is rainwater overflow during time interval t, m^3, L_t is seepage losses during time interval t, m^3.

Considering that closed tanks were adopted in the research, the phenomenon of water evaporation from the tank and direct precipitation onto the water storage surface was omitted in the simulation model. Therefore, the balance equation used in the research took the form:

$$V_t = V_{t-1} + I_t - O_t - Y_t \tag{5.4}$$

The inflow of rainwater to the tank depends on the surface of the roof and precipitation occurring in a time interval. The quantity of runoff I_t during the time step is determined from Eq. (5.5):

$$I_t = \psi \cdot A \cdot R_t \tag{5.5}$$

where ψ is runoff coefficient, A is the roof area, m^2, R_t is rainfall during time interval t, m.

The input parameters of the simulation model used are the following:

- rainwater tank capacity, m^3,
- rainfall during time interval, m,
- demand for non-potable water in time interval per resident, m^3/day,
- number of occupants,
- roof surface, m^2,
- runoff coefficient.

While developing the model, the following main assumptions were also made (Słyś 2009):

- the rainwater tank capacity is fixed,
- demand for non-potable water is satisfied primarily by storaged rainwater, and only then by water from the network,
- any excess of rainwater is drained to the sewage system or to another rainwater management equipment,
- the effect of wind direction and strength and air temperature and humidity,
- the amount of rainwater flowing into the tank depends on the roof surface, the type of roofing and slope.

The model does not take into account the impact of rainwater quality on the functioning of rainwater harvesting systems, but this assumption is the same as other models (Palla et al. 2012; Silva et al. 2015). In addition, rainwater flowing down from the roof of the building is characterized by a low degree of pollution and most often it is sufficient to use a filtration process to pretreat it. An exception is the case in which RWHS laundering is used. Then more advanced rainwater treatment processes are required.

In the research, it was assumed that rainwater would be used only for non-potable uses. It was assumed that the harvested rainwater would be used for toilet flushing, washing, and garden watering. The end use of rainwater was considered in three different options:

Option 1: toilet flushing (Fig. 5.3),
Option 2: toilet flushing and washing (Fig. 5.4),
Option 3: toilet flushing, washing, and garden watering (Fig. 5.5).

Daily demand for non-potable water was calculated based on the number of residents, garden area, and specific water consumption for individual purposes. The demand for each scenario was determined from Eqs. (5.6), (5.7), and (5.8).

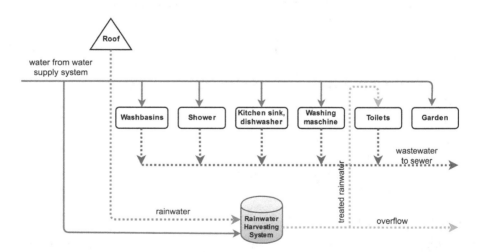

Fig. 5.3 Diagram of system operation in Option 1

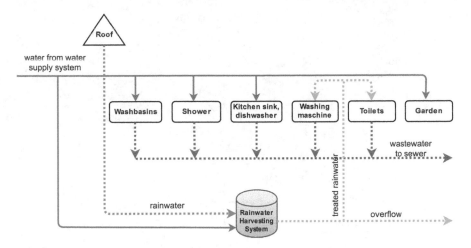

Fig. 5.4 Diagram of system operation in Option 2

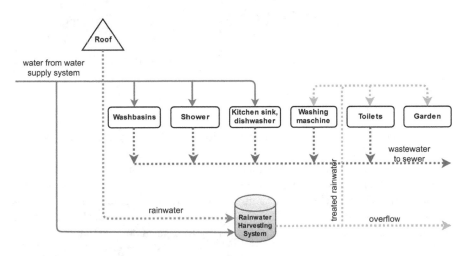

Fig. 5.5 Diagram of system operation in Option 3

$$D_{t1} = O_c \cdot q_t \tag{5.6}$$

$$D_{t2} = O_c \cdot q_t + O_c \cdot q_w \tag{5.7}$$

$$D_{t3} = O_c \cdot q_t + O_c \cdot q_w + G_a \cdot q_g \tag{5.8}$$

where O_c is the number of occupants, q_t is daily water consumption for toilet flushing, dm^3/person/day, q_w is daily water consumption in washing machines,

Fig. 5.6 Location of case study cities in Europe

dm^3/person/day, q_g is daily water consumption for garden watering, dm^3/m^2/day, G_a is garden surface, m^2.

In order to determine the optimal tank capacity, the volumetric reliability V_r of the series of rainwater tanks which were assumed in the first step was determined. The tank capacity was considered optimal when a further increase in this capacity resulted in changes in volumetric reliability of 1% or less. V_r was calculated based on Eq. (5.9), dividing long-term daily supply by long-term daily demand.

$$V_r = \frac{\sum_{t=1}^{T} Y_t}{\sum_{t=1}^{T} D_t} \times 100 \tag{5.9}$$

where V_r is the volumetric reliability of RWHS, %, Y_t is the yield from the store during time interval t, m^3, D_t is the water demand during time interval t, m^3.

Research on the efficiency of using the rainwater harvesting system was carried out at eight different locations in Europe, shown in Fig. 5.6.

The simulation tests were carried out using real archive daily rainfall data, which were recorded at meteorological stations in the analyzed cities. The research used data from 2003 to 2012. The average annual rainfall totals for this period are shown in Table 5.1. According to many researchers, a 10-year rainfall series leads to similar

Table 5.1 Average height of annual rainfall R for the locations analyzed in the years 2003–2012

City/Country	Rainfall R (mm)										Average
	2003	2004	2005	2006	2007	2008	2009	2010	2011	2012	
Prague (The Czech Republic)	309	477	486	480	487	476	646	452	524	672	501
Budapest (Hungary)	838	688	711	515	295	784	831	370	614	670	631
Rome (Italy)	635	679	322	673	740	894	539	656	700	1033	687
Warsaw (Poland)	545	519	514	482	593	547	653	789	601	537	578
Lisbon (Portugal)	850	446	853	375	672	665	1029	741	581	719	693
Bratislava (Slovakia)	326	529	536	568	557	573	583	770	475	561	548
Madrid (Spain)	414	466	423	359	486	290	424	314	213	271	366
Stockholm (Sweden)	545	658	536	564	430	646	614	523	470	502	549

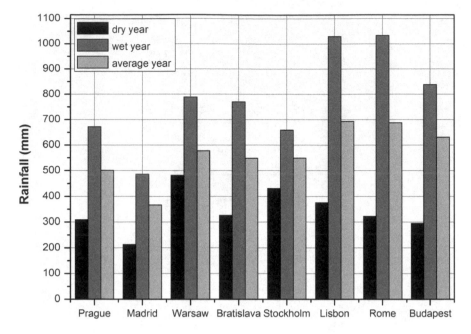

Fig. 5.7 Annual rainfall for the dry and wet year for the locations analyzed

results as for longer rainfall time series, so it was decided to accept such a period for research (Mitchell 2007; Ghisi et al. 2012).

On the basis of rainfall data presented in Table 5.1, it is possible to determine for each location the year with the lowest rainfall sum and the year with the highest rainfall sum (wet year). The data in relation to average rainfall values are shown in Fig. 5.7. As we can see in this figure, the sums of annual rainfall have fluctuated considerably over 10 years. Therefore, it is very important to determine the effectiveness of the rainwater harvesting system, as was done in the research, for the amount of rainfall from a long period of time.

RWHS efficiency is influenced by several parameters such as roof area, demand for non-potable water, rainfall volume, and tank capacity (Lopes et al. 2017). All these factors were taken into account in the research. It was assumed that rainwater collected from the building's roof surface would be transported to the tank. Then the rainwater stored in the tank will be collected by the pump system for the sanitary installation in the building.

The analysis of RWHS functioning was performed using the data listed in Table 5.2. The focus was on single-family buildings because according to Eurostat more than 33% of people live in this type of buildings in Europe. In some of the countries concerned, such as Slovakia, Poland, and Hungary, more than 50% of residents live in single-family buildings (Eurostat 2016). Taking this into account, the input data for the simulation model characterizing such building objects was adopted. The roof surface in single-family buildings usually ranges between 100 and

Table 5.2 Input data of the simulation model characterizing the objects examined

Parameter	Value
Roof area A, m^2	100, 150, 200
Garden area G_a, m^2	250, 500, 750
Number of residents O_c	2, 3, 4
Daily water consumption for toilet flushing q_t dm^3/day/person	35
Daily water consumption in washing machine q_w, dm^3/day/person	16
Daily water consumption for garden watering q_g, dm^3/day/m^2	2.5 (Rome, Madrid, Lisbon)
	1.25 (Budapest, Bratislava, Prague, Warsaw, Stockholm)
Runoff coefficient ψ	0.9
Retention tank volume V, m^3	From 1 to 30

200 m^2, hence the various roof sizes were taken into account in the calculations. To address the losses due to evaporation, leakage, and spilling, a runoff coefficient of the rooftop of 0.9 was used (Słyś 2013). The reliability calculations were a performer for a common storage tank of size ranging from 1 to 30 m^3. The variable parameter of the model was the demand for water of lower quality resulting from the adopted installation variants.

The frequency and amount of water used for watering mainly depend on soil properties and climate conditions (Zhang et al. 2018). By analyzing the climatic conditions and available literature data, the frequency of garden watering, unit water demand for this purpose, and the period of the year when this treatment is performed were determined for each location (Liuzzo et al. 2016; Domenech and Sauri 2011; RMI 2002). It was assumed that in Rome, Madrid, and Lisbon the garden will be watered in May, June, July, August, and September, while for other locations in a similar period except for May.

5.2 The Efficiency of Rainwater Harvesting System

The tests carried out on a simulation model allowed determining the efficiency of the rainwater harvesting system (RWHS) in single-family buildings for their locations in Europe and for assumed options of rainwater use. Their main purpose was to determine volumetric reliability V_r for various tank capacities, and this allowed the determination of the optimal tank size for the changing demand for lower quality water. The V_r parameter not only determines the efficiency of the tank for its various capacities but also real savings of tap water that can be obtained by using RWHS.

Option 1, assuming the use of rainwater only for toilet flushing, was characterized by a daily demand for this purpose in the range from 70 dm^3 (2 persons) to 140 dm^3

(4 persons), depending on the number of residents. Based on the results obtained, it was found that the highest efficiency of RWHS, at 99%, was achieved in Rome. The data presented in Fig. 5.8 show the influence of the roof surface on the V_r value. The larger roof surface increases the amount of rainwater flowing into the tank and thus allows greater water demand for flushing toilets. The difference is not noticeable when two people use the installation and the water demand is low. In this situation, volumetric reliability is not dependent on the size of the roof, and the optimal tank capacity is 7 m^3. A roof of 100 m^2 allows V_r to be 95 and 87%, for 3 and 4 people, respectively. The larger roof area increases volumetric reliability to 99%, but for much larger tank volumes of 13 m^3 (for 3 people) and 19 m^3 (for 4 people).

A similar trend in the charts was obtained for the rainwater harvesting system located in Lisbon. The increase in tank capacity results in a rapid increase in volumetric reliability in the range from 46% (tank capacity 1 m^3) to 98% (tank capacity 7 m^3), which is shown in Fig. 5.9. As the results of the research have shown, further increase of the tank capacity does not significantly increase its efficiency. The impact of the roof surface on V_r, when two people use the installation, is virtually unnoticeable, as was the case for Rome. However, it is particularly visible when the demand for water for toilet flushing was 140 dm^3/day (4 persons). RWHS efficiency for a 200 m^2 roof increased by 11% compared to the smallest roof.

In the case of Madrid, the maximum volumetric reliability was 91% for a 9 m^3 tank (for a roof with an area of 200 m^2 and a water demand for toilet flushing of 70 dm^3/day). With the same surface and higher water demand, the RWHS efficiency was 2–4% lower (Fig. 5.10). The largest differences in the V_r value for different water needs are observed for a 100 m^2 roof. When 4 people use the installation, the V_r was 64% (tank capacity 15 m^3), and if 2 people the efficiency increased to 88% with a smaller tank capacity of 9 m^3.

A slightly lower rainwater harvesting system efficiency of up to 84% was obtained for the RWHS location in Budapest (Fig. 5.11). In contrast to the cases described above, the differences between the level of volumetric reliability for different water needs and different roof sizes were small or nonexistent (Fig. 5.11c). Maximum V_r was achieved for much smaller tank volumes than in Rome, Lisbon, and Madrid. It was 3 m^3, 5 m^3, and 9 m^3 for 2 persons, 3 persons, and 4 persons, respectively.

The same maximum V_r value as for RWHS located in Budapest was obtained for its location in Prague, but this level was achieved for larger tank capacities (Fig. 5.12). The differences between volumetric reliability for different needs of rainwater used for toilet flushing are also more noticeable. For this location of the RWHS system and the calculation parameters adopted, the optimal tank capacity was 5 m^3 for two users of the installation (regardless of the roof surface), 9 m^3 for three people (roof 100 and 150 m^2), and 13 m^3 for four residents (roof 100 and 150 m^2). For the largest roof area, the highest system efficiency of 83% was obtained if the tank volume was 7 m^3 and 11 m^3, for 3 and 4 people, respectively.

When analyzing the test results for rainwater harvesting systems located in Warsaw (Fig. 5.13), Stockholm (Fig. 5.14), and Bratislava (Fig. 5.15), no significant differences were found. For these locations, the maximum value of volumetric reliability was 80%. There were also no significant differences between V_r for different

Fig. 5.8 Efficiency of rainwater harvesting system in Rome, **a** roof surface 100 m², **b** roof surface 150 m², **c** roof surface 200 m²

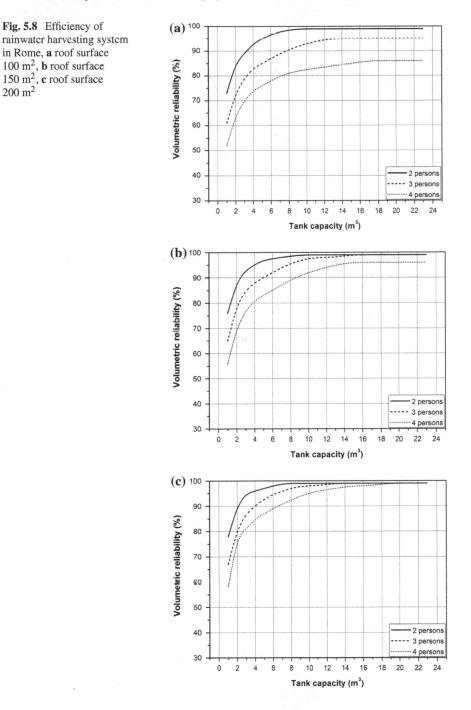

Fig. 5.9 Efficiency of
rainwater harvesting system
in Lisbon, **a** roof surface
100 m², **b** roof surface
150 m², **c** roof surface
200 m²

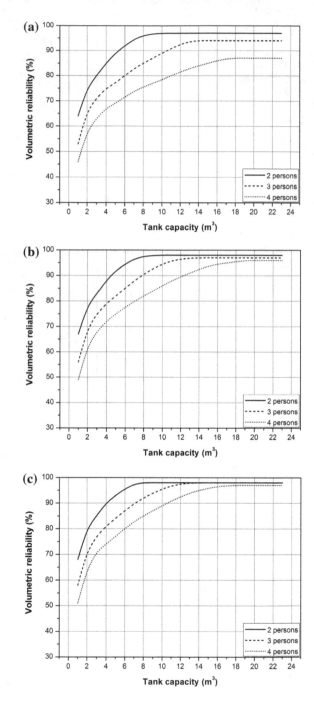

Fig. 5.10 Efficiency of
rainwater harvesting system
in Madrid, **a** roof surface
100 m², **b** roof surface
150 m², **c** roof surface
200 m²

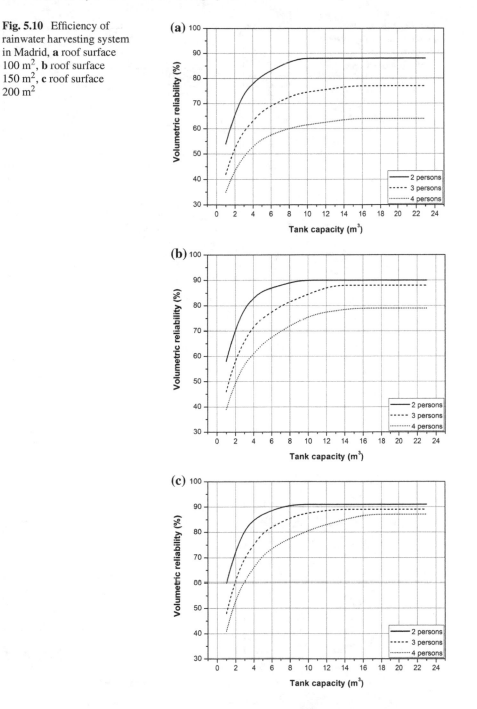

Fig. 5.11 Efficiency of rainwater harvesting system in Budapest, **a** roof surface 100 m^2, **b** roof surface 150 m^2, **c** roof surface 200 m^2

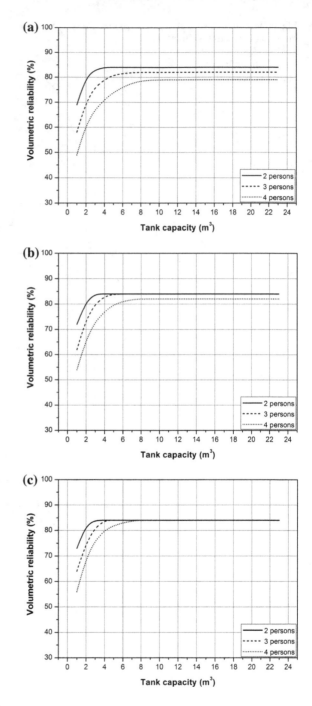

Fig. 5.12 Efficiency of rainwater harvesting system in Prague, **a** roof surface 100 m², **b** roof surface 150 m², **c** roof surface 200 m²

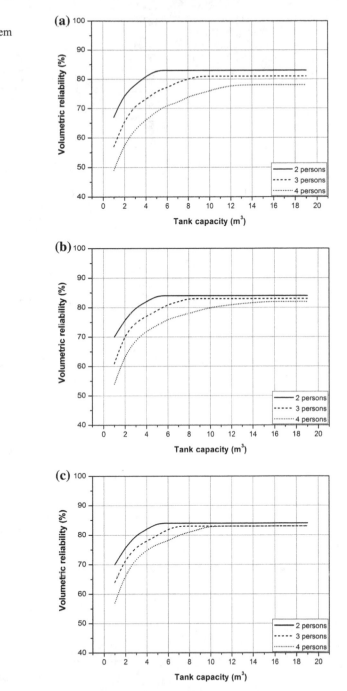

Fig. 5.13 Efficiency of
rainwater harvesting system
in Warsaw, **a** roof surface
100 m², **b** roof surface
150 m², **c** roof surface
200 m²

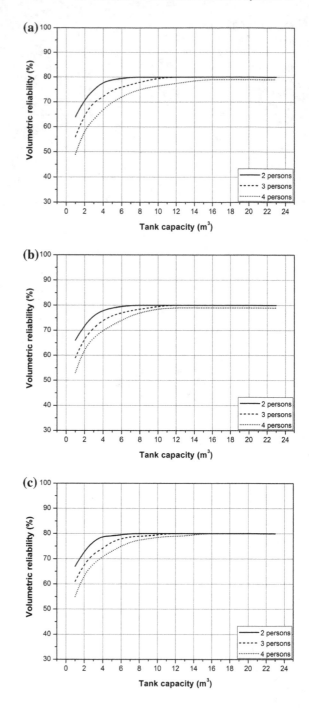

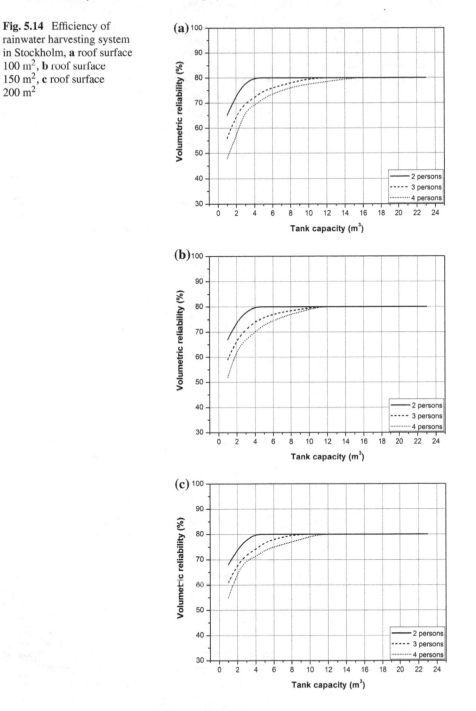

Fig. 5.14 Efficiency of rainwater harvesting system in Stockholm, **a** roof surface 100 m², **b** roof surface 150 m², **c** roof surface 200 m²

Fig. 5.15 Efficiency of rainwater harvesting system in Bratislava, **a** roof surface 100 m², **b** roof surface 150 m², **c** roof surface 200 m²

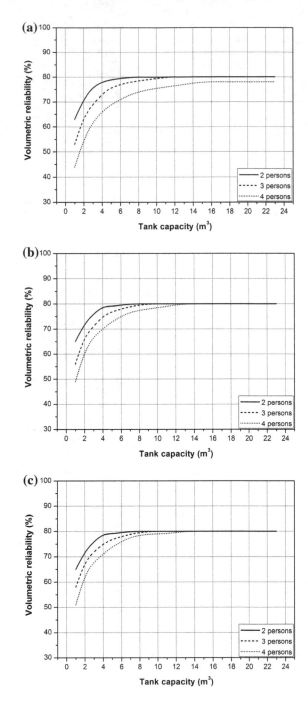

rainwater needs. In the case of Warsaw and Bratislava for demand $q_t = 70$ dm^3/day (2 persons), the highest efficiency of RWHS was ensured by a tank with a capacity of 7 m^3, and for $q_t = 105$ dm^3/day (3 persons) a tank with capacity of 11 m^3 (roof 100 m^2) or 9 m^3 (roof 150 and 200 m^2). In the case of the RWHS location in Stockholm and the water consumption for toilet flushing by two people, the maximum V_r was obtained for a 5 m^3 tank regardless of the roof surface. Whereas for 4 people the optimal tank capacity is 15 m^3 with the famous rainwater from the smallest roof or 11 m^3 for larger roof areas.

Considering the research results for all analyzed locations in Europe, it was found that these results were affected by the length of the period during the year during which rainwater flows into the reservoir and the amount of annual rainfall. This is due to the climatic conditions prevailing in various European countries. The highest volumetric reliability value was characteristic of rainwater harvesting system located in Rome (99%) and Lisbon (98%), where due to warm winters rainwater flows into the reservoir practically all year round and only a small part of it is discharged into the sewage system. The average annual rainfall was the highest for both locations (Table 5.1). In the case of Madrid, despite the fact that rainfall was the lowest for 10 years, the efficiency of RWHS was around 90%. Such a high V_r value was mainly due to the rainwater flowing into the rainwater harvesting system for about 10 months during the year. In other locations, where winter can last for several months, and annual precipitation is at a similar level, differences in volumetric reliability were insignificant.

The second option of rainwater harvesting system (Option 2) assumed the use of rainwater for two purposes: toilet flushing and washing. The total daily demand for these purposes was 102 dm^3, 153 dm^3, and 204 dm^3, for 2, 3, and 4 people, respectively. Tests for this variant, as well as for the first variant, were carried out for tanks with a capacity of 1–30 m^3. Based on the results obtained for each location of the rainwater harvesting system, charts were drawn in which the curves had a similar shape as in the scenario of using rainwater for toilet flushing. For the initial capacities of tanks, there was a sharp increase in volumetric reliability, and after reaching its maximum value, no greater differences than 1% were observed.

The highest efficiency of RWHS for Option 2 was obtained for its locations in Rome and Lisbon. Volumetric reliability values for both locations were at a similar level and differed by several percents. The results of the research in this area are shown in Figs. 5.16 and 5.17. Comparing these results with the V_r values obtained for the first variant of RWHS, it was found that the largest decrease in volumetric reliability is observed when rainwater flows from a 100 m^2 roof. This is due to the lower amount of rainwater flowing into the tank and the increased demand for rainwater in Option 2. The larger the roof, the smaller the V_r drop in Option 2 compared to Option 1. It was also noted that in order to achieve maximum efficiency of RWHS, it is necessary to use larger tank capacities than in the situation when rainwater is used to flush toilets (Option 1).

Similar trends in the results in the second option were also obtained for the rainwater harvesting system located in Madrid (Fig. 5.18). The maximum system efficiency for two people was 77%, 88%, and 89%, for 100 m^2, 150 m^2, and 200 m^2 roof,

Fig. 5.16 Efficiency of rainwater harvesting system located in Rome—Option 2, **a** roof surface 100 m², **b** roof surface 150 m², **c** roof surface 200 m²

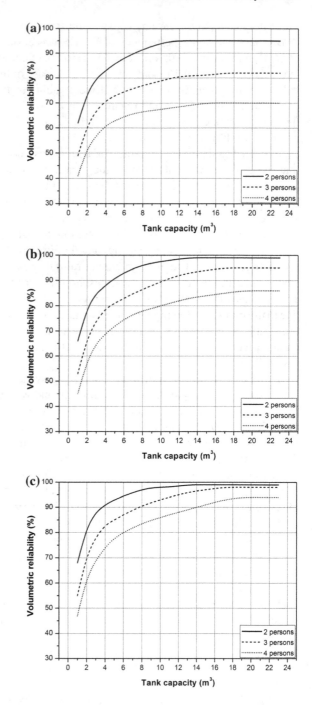

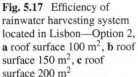

Fig. 5.17 Efficiency of rainwater harvesting system located in Lisbon—Option 2, **a** roof surface 100 m², **b** roof surface 150 m², **c** roof surface 200 m²

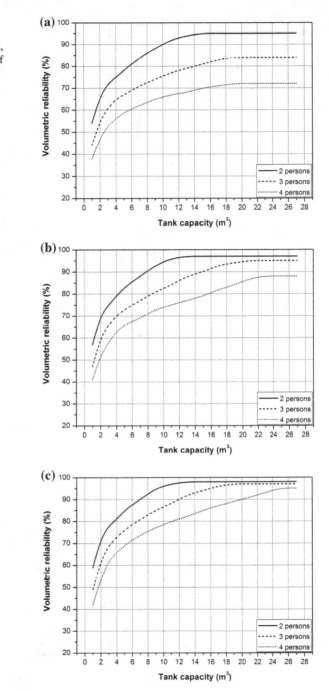

Fig. 5.18 Efficiency of rainwater harvesting system located in Madrid—Option 2, **a** roof surface 100 m², **b** roof surface 150 m², **c** roof surface 200 m²

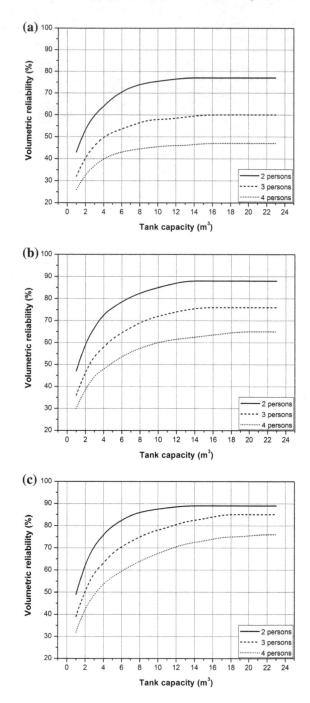

respectively. For a larger number of people and thus a greater demand for water for toilet flushing and washing, volumetric reliability was lower, in some cases by up to 18% (Fig. 5.18a). Comparing the V_r values obtained for this location in Option 1 and Option 2, the smallest differences are observed, as in the case of Rome and Lisbon, for larger roof areas. This is particularly evident when the rainwater consumption is low (2 persons) and the roof area is 150 or 200 m^2.

When analyzing the results for the location of rainwater harvesting systems in Warsaw, Bratislava, and Stockholm, it was noted that the maximum volumetric reliability values obtained in Option 2 were identical to those in Option 1 and were 80%. This is especially visible in cases where two people use the installation. The differences in volumetric reliability between the two variants are noticeable for 3 and 4 people, which is due to the greater demand for water in Variant 2. The selected results for these locations are shown in Fig. 5.19.

Considering the research results for Option 2 and all rainwater harvesting system locations analyzed, it can be generally stated that, as for Option 1, the annual rainfall in each city and the length of the period in which rainwater flows from the roof to the tank have the significant impact on the results. Due to the greater demand for water resulting from its use for toilet flushing and washing, in Option 2 lower efficiency of each system was obtained than in Option 1. These differences were not significant especially in places where, due to the climatic conditions prevailing there, rainwater flows into the tank for most or all year round.

The last of the accepted options of using rainwater assumed its use for toilet flushing for washing and watering the garden (Option 3). In this variant, various sizes of watered areas were assumed in the calculations: 250, 500, and 750 m^2. By determining the length of the period of the year in which the gardens are watered in each location and the amount of water used for this purpose, the climate conditions prevailing in the area were taken into account. Simulations on the model were made with the assumption that in Rome, Lisbon, and Madrid the garden will be watered in the amount of 2.5 dm^3/m^2/day for May, June, July, August, and September, while in other locations in the same months, except May. For these locations, based on literature data, the amount of water used for irrigation was determined at 1.25 dm^3/m^2/day. Based on the results obtained, charts were drawn up, however, due to their large number, selected ones were presented in the monograph. Generally, they did not observe such a sharp upward trend in the value of volumetric reliability, as it was visible in the first and second variants of the rainwater harvesting system.

Unlike Option 1 and Option 2 in Option 3, rainwater harvesting systems located in Rome, Lisbon, and Madrid had the lowest efficiency. The lowest volumetric reliability was obtained for Madrid, where the lowest rainfall occurred during the year. For two system users, the maximum V_r was 29%, 37%, and 42%, respectively, for a roof area of 100 m^2, 150 m^2, and 200 m^2 (Fig. 5.20). For 3 and 4 people, the decrease in volumetric reliability compared to the V_r level for 2 people was insignificant, in the range of 2–4%. The location of RWHS in Rome (Fig. 5.21) and Lisbon (Fig. 5.22) would make it possible to achieve its efficiency at the level of about 50%. The impact of the number of people on the final V_r was also small.

Fig. 5.19 Efficiency of
rainwater harvesting
system—Option 2, **a** RWHS
located in Warsaw, roof
surface 100 m², **b** RWHS
located in Bratislava, roof
surface 150 m², **c** RWHS
located in Stockholm, roof
surface 200 m²

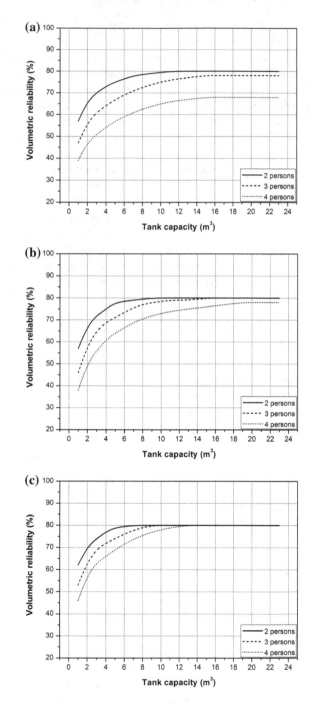

Fig. 5.20 Efficiency of rainwater harvesting system located in Madrid for two users—Option 3, **a** roof surface 100 m², **b** roof surface 150 m², **c** roof surface 200 m²

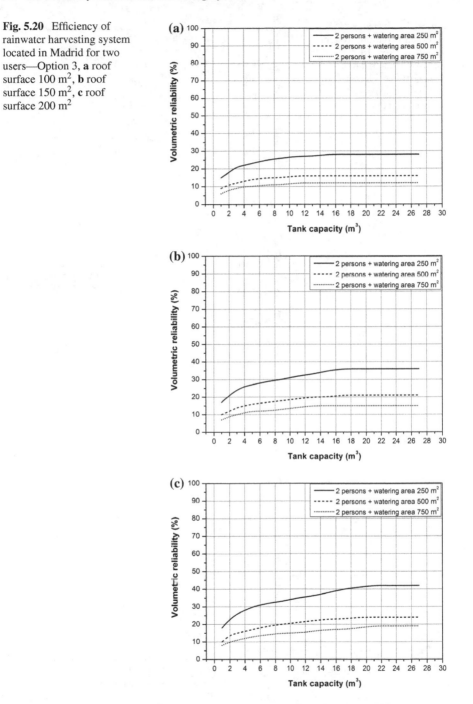

Fig. 5.21 Efficiency of
rainwater harvesting system
located in Rome for three
users—Option 3, **a** roof
surface 100 m², **b** roof
surface 150 m², **c** roof
surface 200 m²

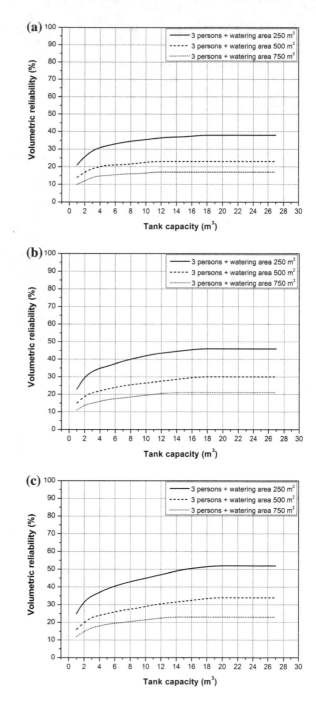

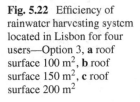

Fig. 5.22 Efficiency of rainwater harvesting system located in Lisbon for four users—Option 3, **a** roof surface 100 m², **b** roof surface 150 m², **c** roof surface 200 m²

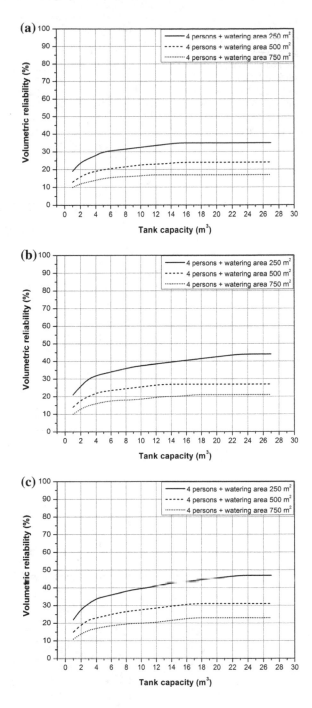

Fig. 5.23 Efficiency of
rainwater harvesting
system—Option 3, **a** RWHS
located in Warsaw, roof
surface 100 m², **b** RWHS
located in Bratislava, roof
surface 150 m², **c** RWHS
located in Stockholm, roof
surface 200 m²

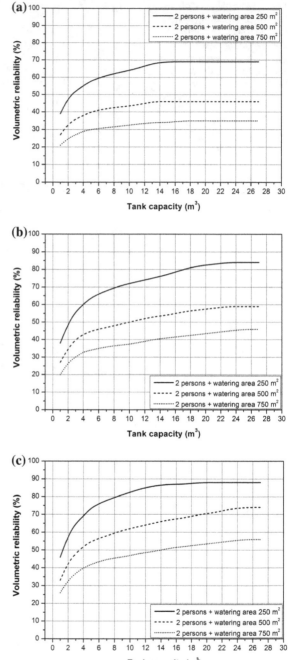

Fig. 5.24 Efficiency of rainwater harvesting system—Option 3, **a** RWHS located in Budapest, roof surface 100 m², **b** RWHS located in Prague, roof surface 100 m²

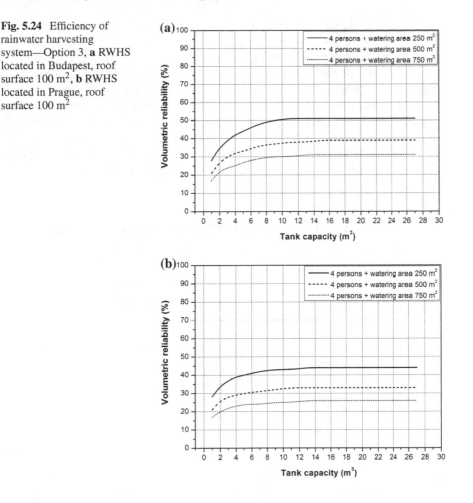

Very similar test results were obtained for other RWHS locations in Warsaw, Bratislava, Stockholm, Prague, and Budapest. The maximum volumetric reliability for these cities was around 90% for the largest roof area. In the case of this installation variant, the impact of the roof surface on the efficiency of RWHS was noticeable and, for example, for a 100 m² roof and 2 system users, the V_r decrease was even 20%. The number of residents and the associated demand for rainwater affected volumetric reliability more than the location of the system in Rome, Lisbon, and Madrid. This was due to the fact that the amount of water for garden watering was smaller than in the case of these three locations and constituted a smaller share in the total daily water demand in Option 3. Figures 5.23 and 5.24 show selected test results for Option 3. As one can see in the graphs the increase in volumetric reliability for these five locations was more rapid than for rainwater harvesting systems located in Rome, Lisbon, and Madrid. Also, the optimal tank volumes for which the highest V_r values were obtained were smaller than it was for RWHS located in these three cities.

When analyzing the results of simulation tests obtained for Option 3, it was found that they had a decisive influence on the water demand for watering the garden. It resulted from the fact that the amount of water used for this purpose, depending on the garden area and the number of residents, constituted from 75% to 86% (garden 250 m^2), from 86% to 92% (garden 500 m^2), and from 90% to 95% (garden 750 m^2) of total rainwater demand in Variant 3 for RWHS locations in Lisbon, Rome, and Madrid. For other locations, this share ranged from 61% to 75%, from 75% to 86% and from 82% to 90%, for a 250 m^2, 500 m^2, and 750 m^2 roof, respectively. This impact was particularly visible in locations where due to the climatic conditions there, a significant amount of water is used for watering during the year.

5.3   The Impact of Rainfall on Volumetric Reliability—Dry Year and Wet Year

In order to analyze in detail the impact of climatic conditions on the efficiency of the rainwater harvesting system, the results of calculated volumetric reliability for years with different rainfall heights were extracted. By analyzing the rainfall data for a 10-year period, the year with the lowest rainfall (dry year) and the year with the highest rainfall (wet year) were determined for each location. Among all RWHS locations, the lowest precipitation occurred in 2011 in Madrid, and the highest in 2012 in Rome (Table 5.1). Due to the extensive data, three RWHS locations in Madrid, Rome, and Stockholm were analyzed. The first two are cities with extreme rainfall levels, while Stockholm has rainfall at a similar level to other locations. Figure 5.24 shows the test results for the three variants of the rainwater harvesting system located in Madrid for the case when the installation is used by two people. On their basis, it was found that there were very large differences in volumetric reliability values between dry and wet years. The largest discrepancies are observed for a 100 m^2 roof, because the runoff of rainwater from its surface is the smallest and the amount of rainfall during the year is decisive for the amount of water flowing into the reservoir, and thus for the overall efficiency of RWHS. The difference in volumetric reliability between dry and wet year is in this case as much as 33% for Option 1, 43% for Option 2 and 18% for Option 3. In the last variant, due to the high demand for rainwater, which is not able to cover the rainwater harvesting system, these differences are smaller. For larger roof areas, the impact of a dry and wet year on RWHS efficiency is less noticeable. This is especially visible in Option 1, where rainwater was intended only for toilet flushing. For this variant, the V_r level in the dry year was only 7% lower compared to the wet year. With the increase in demand for rainwater, the difference in efficiency of RWHS increased and amounted to 18% in Option 2 and 3.

Considering the test results obtained for the rainwater harvesting system located in Rome (Fig. 5.26), it can be seen that the impact of the dry and wet year on the value of volumetric reliability was much smaller than in the case of RWHS in Madrid, especially in Option 1 and Option 2 installations. For these variants, the

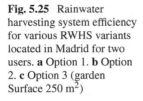

Fig. 5.25 Rainwater harvesting system efficiency for various RWHS variants located in Madrid for two users. **a** Option 1. **b** Option 2. **c** Option 3 (garden Surface 250 m²)

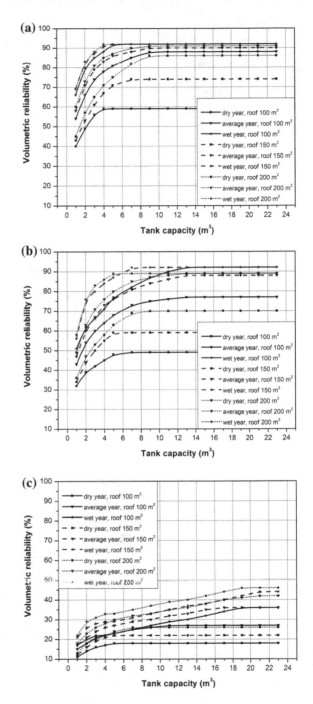

Fig. 5.26 Rainwater harvesting system efficiency for various RWHS variants located in Rome for two users. **a** Option 1. **b** Option 2. **c** Option 3 (garden Surface 250 m²)

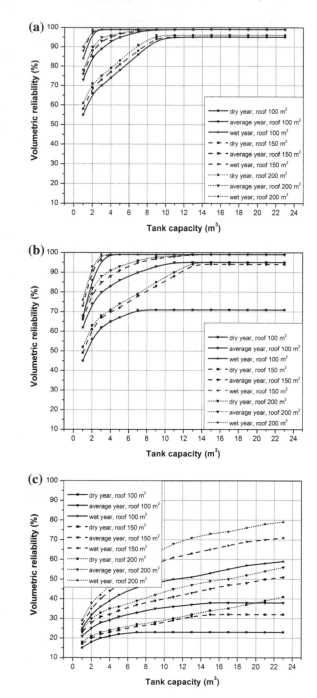

Fig. 5.26 Rainwater harvesting system efficiency for various RWHS variants located in Rome for two users. **a** Option 1. **b** Option 2. **c** Option 3 (garden Surface 250 m²)

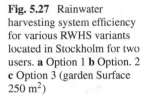

Fig. 5.27 Rainwater harvesting system efficiency for various RWHS variants located in Stockholm for two users. **a** Option 1 **b** Option. 2 **c** Option 3 (garden Surface 250 m²)

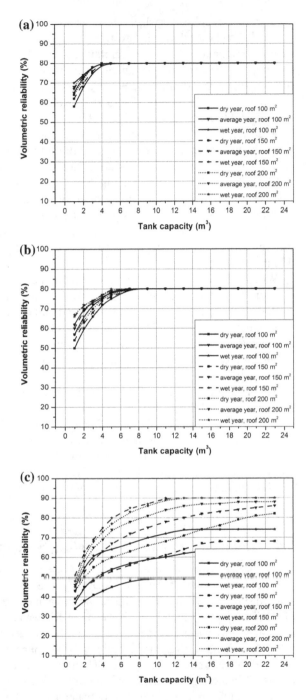

biggest differences occur when rainwater flows to the RWHS from the smallest roof ($100\,\text{m}^2$). The larger the roof, the differences in V_r are less noticeable between the year with the lowest and the highest rainfall. Unlike Madrid, dependencies for Option 3 are shaped. The impact of dry and wet year is more visible here, and the differences between them are significant. For example, for a $100\,\text{m}^2$ roof, the difference in volumetric reliability is 36% (Fig. 5.26c), while for the same parameters, for RWHS in Madrid it was half as much (Fig. 5.25c).

Research in this area was also carried out for a location that was characterized by slight differences in the amount of rainfall in individual years. In Stockholm, the year with the lowest rainfall was 2007, and the highest was 2004. Test results for this RWHS location are shown in Fig. 5.27. It was found that in this case, the impact of dry and wet year on volumetric reliability was only noticeable for Option 3 with the highest water demand. No significant changes were observed for the first two RWHS variants and the maximum V_r level was the same as for the average year. The differences in V_r value in Option 3 are 24%, 22%, and 8% for a $100\,\text{m}^2$, $150\,\text{m}^2$, and $200\,\text{m}^2$ roof, respectively.

When analyzing the above research results, it can be concluded that the impact of a dry and wet year on the efficiency of the rainwater harvesting system is particularly visible for locations with a hot climate and mild winters (Madrid, Rome), where rainfall flows from the roof to the tank throughout the year. In the case of areas with a moderate climate, where over many years there are no large differences in the sum of annual precipitation, and RWHS in winter, due to negative temperatures do not function, this effect is insignificant.

The simulation tests carried out on the rainwater harvesting system model for selected cities in Europe showed how important it was to take into account local climate conditions in the calculation of RWHS efficiency. However, taking into account the financial criterion applicable in most investment cases, it is necessary to carry out a comprehensive analysis of the rainwater harvesting system efficiency, taking into account the economic efficiency of each system. Therefore, such studies were carried out, the effects of which are presented in Chap. 6.

5.4   Model of Graywater Recycling System

The implementation of the graywater recycling system (GWRS) requires a dual sewage system in the building, separating sewage into black and gray. The efficiency of GWRS mainly depends on the demand for water for non-potable uses and the amount of gray water flowing into the tank. It was assumed in the research that gray water from hand washing and bathing will be collected through separate pipes and directed to the pretreatment system. Then, treated gray water will be used to flush toilets. Considering the public opinion expressed in the surveys, whose results are presented in Chap. 8, the use of gray water for washing was not foreseen. In most cases, the respondents were not interested in this way of saving water for hygiene reasons. Due to the content of some substances in the gray water and their possible

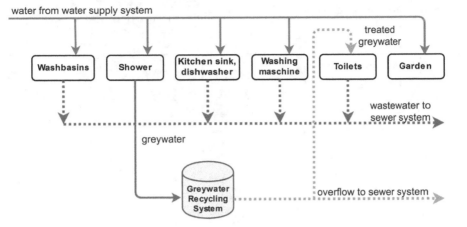

Fig. 5.28 Installation scheme with graywater recycling system located in Rome, Madrid, Lisbon, and Stockholm

adverse effect on the vegetation in the garden, this gray water was not included as a potential source of water in the analyzed buildings. Therefore, GWRS effectiveness tests were carried out for the system layout shown in Figs. 5.28 and 5.29. These systems differ from each other in points from which gray sewage is collected and used. It resulted from different unit water consumptions for individual purposes. The biggest discrepancies for the analyzed locations in Europe occur in the use of water for body washing. The water demand for this purpose is between 26% and 40%. Taking into account the daily water consumption in residential buildings discussed in Chap. 3, the tests were carried out using the data shown in Table 5.3. Depending on the location of the system, the following equations were used to determine the gray water yield Y_{Gt}:

$$Y_{Gt1} = O_c \cdot q_s \tag{5.10}$$

$$Y_{Gt2} = O_c \cdot q_s + O_c \cdot q_h \tag{5.11}$$

where Y_{Gt1} is the graywater yield for GWRS located in Madrid, Lisbon, Rome and Stockholm, dm³/day, Y_{Gt2} is the graywater yield for GWRS located in Warsaw, Bratislava, Budapest and Prague, dm³/day, O_c is number of occupants, q_s is daily water consumption for showering or bathing, dm³/person/day, q_h is daily water consumption for hand washing, dm³/m²/day.

The daily demand for water as well as the graywater inflow to the tank results from the number of residents and the unit water consumption for body and hand washing and washing. These are relatively constant values, which undergo significant changes in certain situations. The demand for treated gray water was determined depending on the installation variant from the formula (5.12).

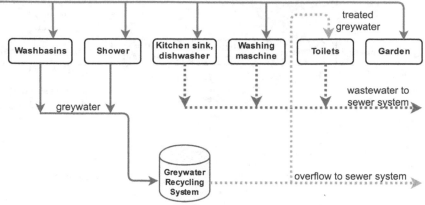

Fig. 5.29 Installation scheme with graywater recycling system located in Warsaw, Prague, Budapest, and Bratislava

Parameter	Value
Number of occupants O_c	2, 3, 4 persons
Daily water consumption for toilet flushing q_t	35 dm³/person/day
Daily water consumption for showering or bathing q_s	30 dm³/person/day (Warszawa, Praga, Bratysława, Budapeszt)
	42 dm³/person/day (Sztokholm, Madryt)
	70 dm³/person/day (Lizbona, Rzym)
Daily water consumption for hand washing q_h	10 dm³/person/day

Table 5.3 Input data for calculating the efficiency of the graywater recycling system

$$D_{Gt1} = O_c \cdot q_t \qquad (5.12)$$

where D_{Gt1} daily water demand for toilet flushing, dm³/day, q_t is daily water consumption for toilet flushing, dm³/person/day.

Depending on the type of graywater recycling system, the optimum storage capacity for treated gray water should be determined taking into account the peak capacity treatment rate and the graywater demand, usage, or behavior patterns (BS 2010). It is recommended that the storage time of treated gray water should be minimized to that needed for immediate use. Storage of untreated gray water should be avoided. Waste-free sewage is usually kept in the tank for no more than one day. Considering these recommendations, it was assumed that the green recycling system in each variant will be selected for the efficiency resulting from the water demand for the given purpose, and the excess gray water will be discharged into the sewage system.

Due to the stability of the gray water inflow to the system and the amount of it used in the variants analyzed, it was not necessary to carry out simulation tests and calculations of the hydraulic efficiency of these systems, as was done for the rainwater harvesting system. Therefore, graywater recycling systems were selected for constant system performance resulting from the demand for non-potable water. Financial analysis was carried out for these systems, and its results are presented in Chap. 6.

References

BS 8525-1:2010, Grey water systems, Part 1: code of practice

Domènech L, Saurí D (2011) A comparative appraisal of the use of rainwater harvesting in single and multi-family buildings of the Metropolitan Area of Barcelona (Spain): social experience, drinking water savings and economic costs. J Clean Prod 19:598–608

EN 16941-1:2018, On-site non-potable water systems—Part 1: systems for the use of rainwater. European Committee for Standardization, Brussels

Eurostat (2016) https://ec.europa.eu/eurostat/statistics-explained/index.php/Living_conditions_in_Europe_-_housing_quality

Fewkes A (2000) Modelling the performance of rainwater collection systems: towards a generalised approach. Urban Water 1:323–333

Ghisi E, Cardoso KA, Rupp RF (2012) Short-term versus long-term rainfall time series in the assessment of potable water savings by using rainwater in houses. J Environ Manag 100:109–119

Liuzzo L, Notaro V, Freni G (2016) A reliability analysis of a rainfall harvesting system in southern Italy. Water 8:1–20

Lopes V, Marques G, Dornelles F, Medellin-Azuara J (2017) Performance of rainwater harvesting systems under scenarios of non-potable water demand and roof area typologies using a stochastic approach. J Clean Prod 148:304–313

Mitchell VG (2007) How important is the selection of computational analysis method to the accuracy of rainwater tank behaviour modelling? Hydrol Process 21(21):2850–2861

Palla A, Gnecco I, Lanza LG, La Barbera P (2012) Performance analysis of domestic rainwater harvesting systems under various European climate zones. Resour Conserv Recycl 62:71–80

Rozporządzenie Ministra Infrastruktury z dnia 14 stycznia 2002 r. w sprawie przeciętnych norm zużycia wody (Dz. U. 2002 Nr 8, poz. 70)

Silva C, Sousa V, Carvalho N (2015) Evaluation of rainwater harvesting in Portugal: application to single-family residences. Resour Conserv Recycl 94:21–34

Słyś D (2009) Potential of rainwater utilization in residential housing in Poland. Water Environ J 23:318–325

Słyś D (2013) Zrównoważone systemy odwodnienia miast. Dolnosląskie Wydawnictwo Edukacyjne, Wrocław

Zhang S, Zhang J, Jing X, Wang Y, Wang Y, Yue T (2018) Water saving efficiency and reliability of rainwater harvesting systems in the context of climate change. J Clean Prod 196:1341–1355

Chapter 6
Research on the Financial Effectiveness of Alternative Water Supply Systems in European Countries

Abstract Currently, the financial criterion is a decisive criterion in the process of making investment decisions. Taking this into account, a financial analysis was carried out for various variants of sanitary installations in single-family buildings located in selected cities in Europe. The research was carried out using the Life Cycle Cost methodology.

6.1 Life Cycle Cost Methodology

There are many different definitions of the Life Cycle Cost (LCC) methodology in the literature. However, regardless of whether they concern the costs of a single product, equipment, entire production technology, or buildings and structures, LCC analysis includes the same main components.

LCC is the sum of costs incurred during the life cycle of a given product and includes capital expenditure, costs of use, and costs of liquidation or economic use (White and Ostwald 1976; Woodward and Demirag 1989; SAE 1999; Kowalski et al. 2007). The value of LCC costs in a general way can be recorded by Eq. (6.1).

$$LCC = INV + OMC + DMC \qquad (6.1)$$

where

INV   investments, €;
OMC operating costs; €;
DMC costs of liquidation or economic use, €

© Springer Nature Switzerland AG 2020
A. Stec, *Sustainable Water Management in Buildings*,
Water Science and Technology Library 90,
https://doi.org/10.1007/978-3-030-35959-1_6

The idea of the Life Cycle Cost methodology was created in the United States of America in the 1960s. The Department of Defense of the United States introduced it to practice in the implementation of public procurement (Epstein 1996). The department also published recommendations for the use of LCC analysis in various areas related to the functioning of the American army (Military Handbook… 1983a; Military Handbook Military Handbook… 1984a, b).

At the same time, there was a sharp increase in the cost of ownership and maintenance of buildings in the United States, which led to the demand to take into account the choice of the possibility of using different construction technologies already at the stage of making investment decisions. In relation with the above in 1978, the American Institute of Architecture published a book where it described the possibilities of using LCC analysis by architects in the design process of construction works (Haviland 1978). In subsequent years, attempts were made to adapt the LCC methodology to related investments with urban infrastructure (Dell Isola and Kirk 1981; Flanagan and Norman 1987; Flanagan et al. 1987).

In the 1980s, the US Department of Energy introduced a statutory requirement to use LCC analysis when purchasing federal facilities, primarily to compare alternative water and energy supply concepts, including renewable energy (Fuller and Petersen 1996).

Currently, LCC cost analysis is used in various fields of the economy, including power engineering, industry, transport, construction, infrastructure, or pumping technology. It is mainly used as a tool in decision-making and management (Bakis et al. 2003; Gluch and Baumann 2004). By carrying out the economic efficiency of a given investment, or the cost-effectiveness of purchasing a product throughout the life cycle, there is an option of choosing the cheapest solution, i.e., the one in which the total costs are minimized (Landers 1996). Often, it happens that the only selection criterion is the initial investment expenditure. This is an easy-to-use criterion but it can lead to a wrong decision in financial and environmental terms, as in some cases, the cost of use may exceed the cost of acquisition several times.

The LCC analysis results can provide valuable information and help make decisions when assessing and comparing alternatives. In many countries, the Life Cycle Cost methodology is legally required for the implementation of new investments, especially those characterized by high initial costs and long service life.

The use of LCC analysis can be important to assess the cost of existence of buildings and structures that have been in use for decades. Taking it into account already at the stage of building design and selection of various materials and technologies may affect not only the initial capital expenditure but above all, the value of operating costs, which are incurred throughout the life of the building.

The use of the LCC technique in the assessment of investment projects related to building development is recommended by the European standard ISO 15686–5 (Buildings… 2008). In 2006, the General Directorate for Entrepreneurship and Industry at the European Commission began the work to formulate a common Life Cycle Cost methodology used in sustainable construction. The effect of these works was, among others, several publications whose aim is to disseminate this methodology in the European Union countries (Life Cycle… 2007a, b, c).

LCC analysis is considered as one of the elements of sustainable development, which is a priority of the "Europe 2020 Strategy" adopted in 2010. According to it, this development is characterized by the support of a resource-efficient economy, a more competitive economy, and an environment-friendly economy.

The Life Cycle Cost costing process consists of several main steps (Fig. 6.1). The detailed tasks (stages) of the LCC analysis may vary and depend on which product or investment it is being carried out. The first stage of the analysis is to determine the problem that LCC analysis is to address and define the main goals of this analysis and its scope. The second stage is the preparation of the cost structure. This stage is divided into cost categories, i.e., capital expenditure and operating costs, which are identified by collecting the correct input data of the model. The next stage involves the development of a comprehensive LCC model, which consists of detailed investment and operating cost models. This model describes reality by mathematically recording the sum of costs associated with the problem under study. As a part of the fourth

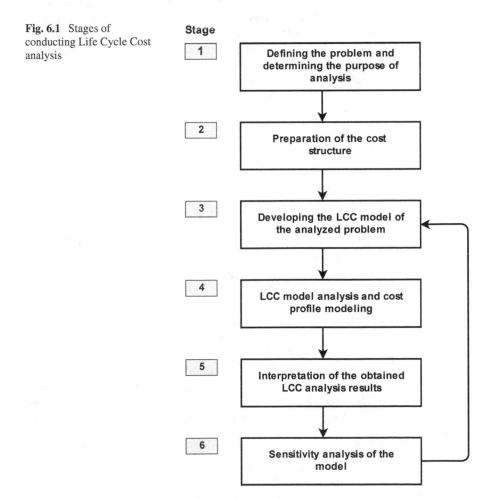

Fig. 6.1 Stages of conducting Life Cycle Cost analysis

stage, calculations of individual costs are made, which constitute the input data for the next stage, which is the interpretation and evaluation of the obtained results. The final stage is the sensitivity analysis. It is carried out in order to examine the impact of changes in model data on LCC, e.g., changes in energy prices over the assumed lifetime of the facility. In the event that the formulated objectives of the analysis are not achieved in the first stage or if errors in the model are found, then it is necessary to introduce some changes in the third stage.

Sensitivity analysis is performed to examine the impact of future changes in the key parameters of the model on the level of LCC costs and, therefore, on the profitability of the investment. Its results allow determining at the decision-making stage, the sensitivity of the model to changes in various variables such as the discount rate, energy prices, and water prices (Pastusiak 2003; Brigham and Ehrhardt 2008; Lang 2007).

In the LCC sensitivity analysis of facilities that will be used for decades, it is important to consider changes in data that affect the value of operating costs (Life Cycle... 2011). The results of such research may help a person who makes an assessment of the profitability of a given investment make a rational decision or by changing the assumptions of the LCC model will allow recalculating the investment project.

The investment sensitivity assessment can be made using various techniques. In the simplest form, the sensitivity analysis involves examining the impact of percent deviations of individual model variables on the value of the decision criterion, for example, LCC costs or operating costs alone. This method assumes a deviation of the variable from its base value, e.g., from -10 to $+10\%$ and again for this data, the total value of the investment is calculated (Rogowski 2004). It is important to accept only one data change at a time to determine how much the LCC cost will change if the value of this variable changes by x%.

The costs of the analyzed investment, which appear in various periods of its use, should not be compared or balanced directly without taking into account the variable value of money over time. Therefore, in the LCC analysis, this is usually taken into account by applying a classical method using the updated value of cash flows net present value (NPV) or present value (PV) (Góralczyk and Kulczycka 2003; Clift 2003; Rogowski 2004; Kowalski et al. 2007; Life Cycle... 2007a). To maintain the requirement of temporary comparability of individual elements included in the cost analysis, a discount account should be used throughout the lifetime of the investment. It allows calculating the value of future expenses or profits and bringing them back to the first base year of the investment, the one in which the decision is made. In line with the guidelines included in the report of the TG4 group, which, on the recommendation of the European Commission, developed guidelines for the use of LCC analysis in construction, LCC costs should be calculated as the current value (PV) of total future expenditures incurred for the operation and use of a building (Task Group... 2003). The present value of the investment can be determined on the basis of the formula (6.2).

$$PV = \sum_{t=1}^{T} \frac{OMC_t}{(1+r)^t} \qquad (6.2)$$

where

PV   the present value of the accumulated annual costs incurred in the future for the use of the building;

OMC_t operating costs in a year t, €;

T   duration of the LCC analysis, years;

r   constant discount rate;

t   the number of years after installation;

Taking the above into account, the final value of LCC costs for each installation variant was determined from the formula (6.3).

$$LCC = INV + \sum_{t=1}^{T} (1+r)^{-t} \cdot OMC_t \qquad (6.3)$$

Guidelines contained in (DOE 2014) studies provide that wherever the life span of an analyzed system goes beyond the foreseeable future, its residual value at the completion of exploitation need not be defined in quantifiable terms. Consequently, such costs were not taken care of in the quantitative analysis, similar to studies conducted by other authors (Rahman et al. 2012).

6.2 Variants of Sustainable Water Management in Buildings

The analysis of the profitability of the application of individual solutions of installations supplied from alternative water sources was carried out for the following system configurations:

- Variant 0—traditional installation supplied with water from the water supply network and sewage discharged into the sewage system (Fig. 6.2),
- Variant 1—rainwater installation for toilets flushing (Fig. 6.3),
- Variant 2—rainwater installation for toilets flushing and washing (Fig. 6.4),
- Variant 3—installation using rainwater for toilets flushing, washing, and watering the garden (Fig. 6.5),
- Variant 4—installation using gray water for toilets flushing (Figs. 6.6 and 6.7),
- Variant 5—installation using rainwater and gray water for toilets flushing, washing, and watering the garden (Figs. 6.8 and 6.9).

Due to the fact that in most of the considered cases, the sum of water demand for flushing toilets and watering the garden significantly exceeds the inflow of gray sewage into the system, it would be unprofitable to use it. Therefore, the financial

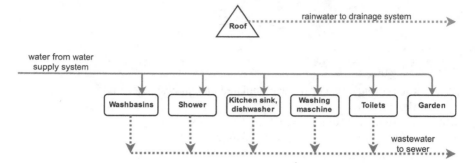

Fig. 6.2 The scheme of a traditional installation supplied with water from a water supply system and sewage discharged to a sanitary and rainwater sewage system—Variant 0

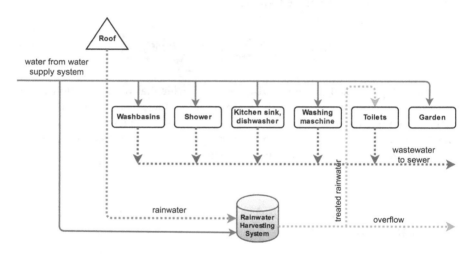

Fig. 6.3 The scheme of the installation supplied with water from a water supply system and additionally equipped with a rainwater harvesting system for toilets flushing and sewage discharged into the sewage system—Variant 1

analysis was carried out for a variant in which gray water was used only for toilets flushing (Variant 4) and a hybrid system (Variant 5), where treated gray sewage is used for flushing toilets and rainwater for washing and watering the garden.

6.3 Case Studies

The results obtained on the simulation model of rainwater harvesting systems (RWHS) and data on the graywater recycling systems (GWRS) presented in Chap. 5 were used as input data to assess the financial effectiveness of the investment consisting of the implementation of these systems in different variants in single-family

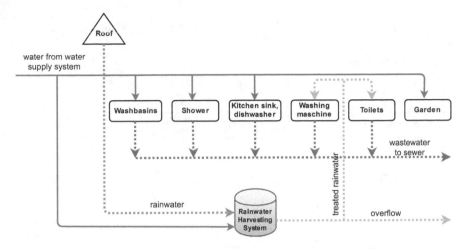

Fig. 6.4 The scheme of the installation supplied with water from a water supply system and additionally equipped with a rainwater harvesting system for toilets flushing and washing, as well as sewage discharged into the sewage system—Variant 2

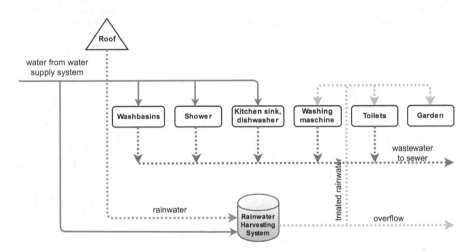

Fig. 6.5 The scheme of the installation supplied with water from a water supply system and additionally equipped with a rainwater harvesting system for toilets flushing, washing, and watering the garden, as well as sewage discharged into the sewage system—Variant 3

houses. The research was carried out for houses located in European cities: Madrid, Lisbon, Rome, Prague, Bratislava, Budapest, Stockholm, and Warsaw.

The studies took into account the purchase and installation costs of water supply and sewage pipes in the building. In addition, in the options assuming the use of alternative water sources, the capital expenditure necessary to bear in connection with the implementation of the described systems was taken into account. In variants

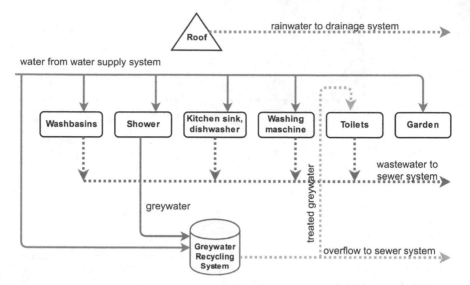

Fig. 6.6 The scheme of the installation supplied with water from a water supply system and additionally equipped with a graywater recycling system for toilets flushing and sewage discharged into the sewage system—Variant 4 located in Stockholm, Madrid, Lisbon, and Rome

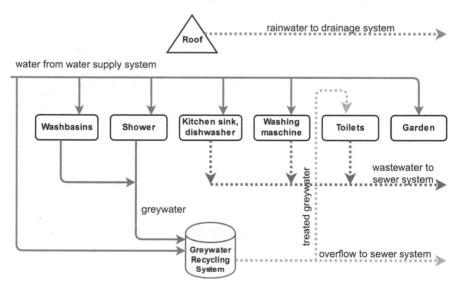

Fig. 6.7 The scheme of the installation supplied with water from a water supply system and additionally equipped with a graywater recycling system for toilets flushing and sewage discharged into the sewage system—Variant 4 located in Warsaw, Prague, Bratislava, and Budapest

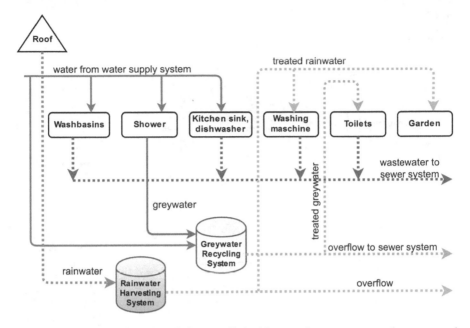

Fig. 6.8 The scheme of the installation supplied with water from a water supply system and additionally equipped with a graywater recycling system for toilets flushing and rainwater harvesting system for washing and watering the garden as well as sewage discharged into the sewage system— Variant 5 located in Stockholm, Madrid, Lisbon, and Rome

1, 2, and 3, it was RWHS, while in variant 4 a, graywater recycling system. In turn, variant 5 includes the purchase cost of both systems.

In all variants, on the basis of the formulated simulation models, the operating costs resulting from water purchase and costs caused by the discharge of sanitary and rainwater sewage into sewage systems were calculated. Unit prices for these services have been established for each location from the information provided by the water supply companies. In configurations with alternative water sources, additional operating costs resulting from the use of electricity for pumping gray water and rainwater from tanks to the installation were also taken into account. In addition, according to the manufacturer's guidelines, these options also include the cost of replacing filters and pumps.

The GWRS adopted for the analysis is a professional system with advanced methods of purification and disinfection that enable the removal of 99.9% of contaminants, viruses, and bacteria. In turn, the cost of RWHS depended on the capacity of the tank. The tests included rainwater harvesting systems with a capacity of 1, 2, 3, 4, 5, 7, 9, and 11 m³ offered by European producers, which are typical for single-family houses. In cases where rainwater was used for washing, additional filters and disinfection were provided.

The discount was assumed to be 5%, as it was used in calculations by (Morales-Pinzon et al. 2012; Roebuck et al. 2011; Rahman et al. 2010; Liaw and Tsai 2004).

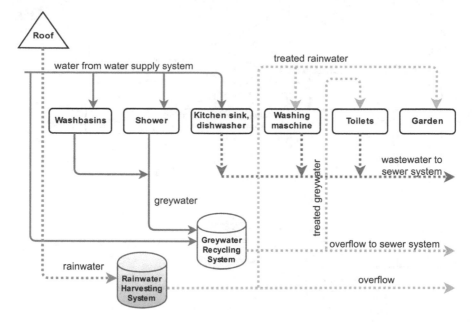

Fig. 6.9 The scheme of the installation supplied with water from a water supply system and additionally equipped with a graywater recycling system for toilets flushing and rainwater harvesting system for washing and watering the garden as well as sewage discharged into the sewage system—Variant 5 located in Warsaw, Prague, Bratislava, and Budapest

Similar to (Morales-Pinzón et al. 2012), the research also included the annual increase in water, sewage, and energy prices. These values were determined based on archival data. Considering the lifetime of materials and equipment currently used, the length of the LCC analysis period has been set at 30 years. Data accepted for calculating LCC costs of the analyzed installation variants in a single-family building are summarized in Tables 6.1 and 6.2. Due to the ever longer periods of drought and the occurring water deficit in Madrid and Lisbon, the annual increase in water prices is significant. In some cities such as Warsaw, Bratislava, or Budapest, water supply companies have set the price for water purchases on a constant level over the next few years. However, when analyzing archival data, a slight increase in this price was determined for these locations. On the basis of Eurostat data, average prices of electricity were adopted and its annual growth. The studies also take into account the costs resulting from discharging rainwater to the sewerage network, but only in those locations where they occur.

Table 6.1 Data used in the calculation of LCC costs

Parameter	Parameter value
Analysis period T	30 years
The cost of filter change in GWRS after each 10 years	€990
The cost of purchasing and installing the GWRS 250 dm³/day INV_{GWHS_250}	€5500
The cost of purchasing and installing the RWHS $INV^3_{RWHS\_1m}$	€2216
The cost of purchasing and installing the RWHS $INV^3_{RWHS\_2m}$	€2259
The cost of purchasing and installing the RWHS $INV^3_{RWHS\_3m}$	€2313
The cost of purchasing and installing the RWHS $INV^3_{RWHS\_4m}$	€2421
The cost of purchasing and installing the RWHS $INV^3_{RWHS\_5m}$	€2569
The cost of purchasing and installing the RWHS $INV^3_{RWHS\_7m}$	€2729
The cost of purchasing and installing the RWHS $INV^3_{RWHS\_9m}$	€3106
The cost of purchasing and installing the RWHS $INV^3_{RWHS\_11m}$	€3647
The cost of purchasing and installing the sanitary systems INV_0	€2500
The discount rate r	5%

Table 6.2 Unitary prices accepted for LCC analysis

City	The cost of purchasing water from the water-pipe network and sanitary sewage discharge to the sewage network in the year 0, €/m³	The annual increase in the prices of purchase of water from the water-pipe network and sanitary sewage discharge to the sewage network, %	The cost of purchasing electricity in the year 0, €/kWh	The annual increase in electricity prices, %	The cost of the discharge of rainwater to the sewage network in the year 0
Bratislava	2.23	2	0.15	2	–
Budapest	2.94	2	0.11		–
Lisbon	2.26	20	0.23		–
Madrid	3.16	12	0.25		–
Prague	3.49	9	0.16		=
Rome	3.50	6	0.22		0,23 €/m³
Stockholm	2.30	4	0.20		38,76 €/year/house
Warsaw	2.31	4	0.14		–

6.4   Analysis Results

6.4.1   Results of Life Cycle Cost Analysis

The research conducted showed that the selection of the right investment option had a decisive impact on the total amount of life cycle costs of the plumbing installation in the residential buildings considered. The results of calculations for the LCC indicator obtained for various conditions of use of the installation indicate that both the number of inhabitants and the price of purchasing water and wastewater disposal determine the profitability of applying individual solutions in the analyzed European cities.

The highest LCC value exceeding several times the value of these costs for other locations was obtained when the alternative installation systems were located in Lisbon (Fig. 6.10). The research results obtained showed that Variant 0 is not an optimal solution for any of the computational cases analyzed. This is due to the fact that the operating costs associated with the operation of a traditional installation solution (Variant 0) over a 30-year period are greater than when using individual installation systems powered from alternative water sources, despite the fact that they require higher investments. Regardless of the number of inhabitants, the most advantageous in financial terms would be to use a variant in which both gray water and rainwater (Variant 5) are used, and the optimal tank capacity in RWHS was 11 m^3. For this capacity of the tank, this variant was characterized by LCC costs lower by almost 43,000 EUR compared to Variant 0, despite the fact that the investment expenditure for the implementation of Variant 5 was five times higher than in the case of a traditional installation solution. Comparing the variants with the use of gray wastewater or rainwater, the latter ones are definitely more favorable because the investments for the implementation of RWHS even with a large capacity tank are lower than those that should be allocated to the installation additionally equipped with a graywater recycling system.

Three times lower LCC values were obtained for the locations of the analyzed systems in Madrid (Fig. 6.11). However, in this case, the traditional installation solution (Variant 0) had the highest LCC costs when the building was inhabited by three or four people. If the installation was used by two users, the least cost-effective variant was Variant 4, in which only the gray water was the alternative source of water. It was observed that, similar to Lisbon for three and four inhabitants, the installation solution where the rainwater harvesting system and the graywater recycling system were implemented was the most financially advantageous. Such a hybrid system allowed achieving the largest water savings in the period of 30 years, which in turn resulted in the lowest operating costs. These benefits were obtained for RWHS equipped with a 9 m^3 tank. The situation looks different with the lowest number of inhabitants. If the installation is used by two people, the most profitable would be to use Variant 3 with a 5 m^3 tank where rainwater is used to flush toilets, wash, and water the garden (Fig. 6.11a). The implementation of RWHS with a larger tank capacity or the addition of a GWRS would only result in an increase in investment outlays that would not be offset by greater water savings.

Fig. 6.10 Results of the
financial analysis for the
location of the building in
Lisbon (roof area 150 m^2,
watering garden area 500 m^2,
a two persons, **b** three
persons, **c** four persons

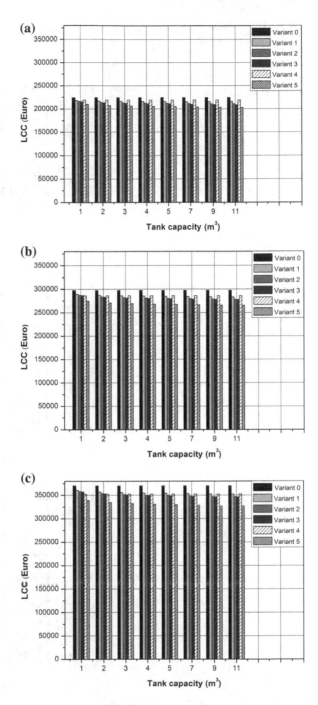

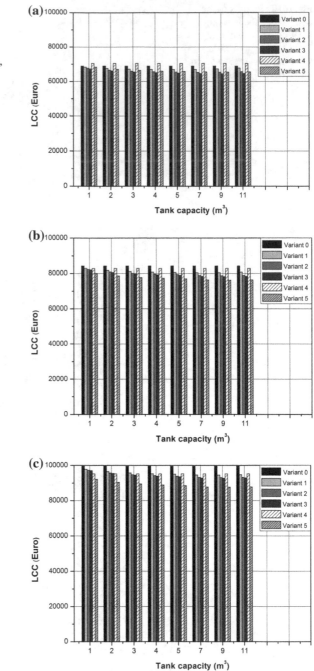

Fig. 6.11 Results of the financial analysis for the location of the building in Madrid (roof area 150 m^2, watering garden area 500 m^2, **a** two persons, **b** three persons, **c** four persons

Also in the case of Prague, it would be beneficial to use alternative water sources in a single-family house (Fig. 6.12). However, for this location, in contrast to Lisbon and Madrid, the most advantageous in financial terms, regardless of the number of inhabitants, was option 3 consisting of the implementation of RWHS. The number of people and the related water demand had an impact on the profitability of using Variant 3 depending on the tank capacity in this system. If the system was used by two people, the Variant 3 with a 5 m³ tank is financially optimal, while for a larger number of inhabitants, it is RWHS equipped with a 7 m³ tank. The highest value of the LCC index for all calculation cases for this location was obtained for variants in which gray water (Variant 4 and Variant 5) constituted an additional source of water.

Variant 3 where rainwater was used to flush toilets, wash, and water the garden also in the case of building location in Rome was the solution with the lowest LCC costs (Fig. 6.13). It was the only city among the eight analyzed that has annual fees for each cubic meter of rainwater discharged into the sewage system. As the results of the research have shown, it is the operating costs associated with the discharge of these waters to the drainage network that have decided about the profitability of using the variant with the rainwater harvesting system. If they were not only for the roof with the largest area (200 m²) and high demand for water (four persons), Variant 3 would still be the most cost-effective option. The differences between the LCC index values for Variant 0 and Variant 3 were insignificant, especially for a roof with a small area determining less rainwater runoff (Fig. 6.13a). The optimal tank that achieves the greatest financial benefits is a tank with a capacity of 7 m³. In only one calculation case (two people, 100 m² roof), the optimal tank capacity is 4 m³. The solution for the installation with the highest LCC costs, regardless of the number of people and the roof area, was also option 4 with gray water for this location.

Very similar results of financial analysis were obtained for the locations of the systems in Budapest (Fig. 6.14a), Bratislava (Fig. 6.14b), Warsaw (Fig. 6.14c), and Stockholm (Fig. 6.14d). The variant with the lowest costs for these locations was the traditional solution of the installation with water supply from the water supply network and sewage discharge to the sewage system (Variant 0). The results and hierarchy of profitability of individual variants were not significantly affected by either the number of users of the installation or the size of the roof area. The largest differences in the LCC ratio were observed when comparing Variant 0 and variants using graywater recycling systems, i.e., Variant 4 and Variant 5. This was due to high capital expenditure that should be incurred when implementing GWRS. The use of alternative sources of water in single-family buildings for these locations was completely unprofitable, as the water savings obtained over a period of 30 years did not cover the capital expenditure and operating costs caused by replacing filters and pumps in both systems. Comparing both alternative water sources, it can be stated that much better financial results were obtained for rainwater harvesting systems.

When analyzing the results of the research in the scope of the LCC analysis for all the locations considered, it was noticed that the use of alternative variants of water installation in single-family buildings is financially beneficial for their locations where the purchase price of water from the water supply network and for sewage disposal was above 3 EUR/m³ (Madrid, Prague, Rome) or if the annual increase

Fig. 6.12 Results of the financial analysis for the location of the building in Prague (roof area 150 m^2, watering garden area 500 m^2), **a** two persons, **b** three persons, **c** four persons

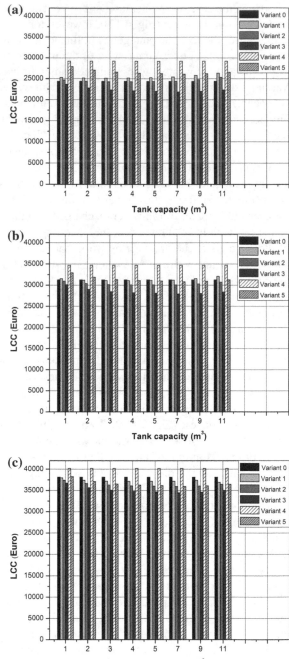

Fig. 6.13 Results of the financial analysis for the location of the building in Rome (500 m² watering garden), **a** two persons, roof area 100 m², **b** three persons, roof area 150 m², **c**) four persons, roof area 200 m²

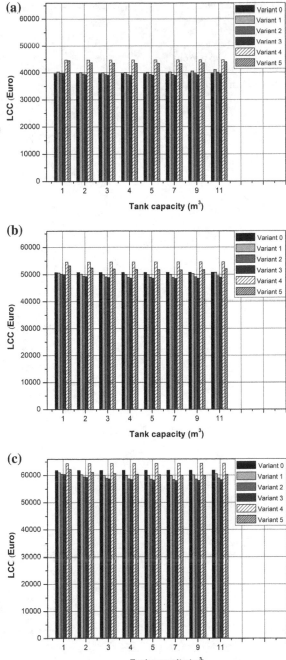

Fig. 6.14 Results of the
financial analysis for the
building location in
a Budapest (two persons),
b Bratislava (two persons),
c Warsaw (three persons),
d Stockholm (four persons)

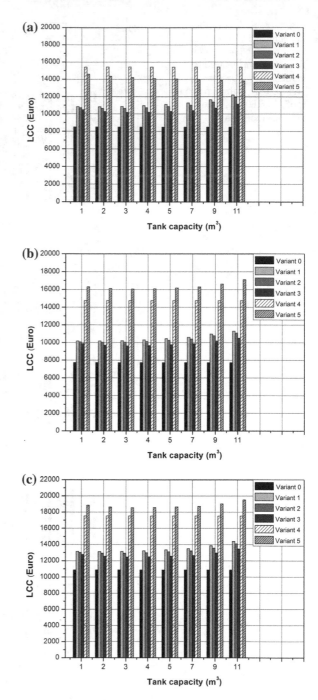

Fig. 6.14 (continued)

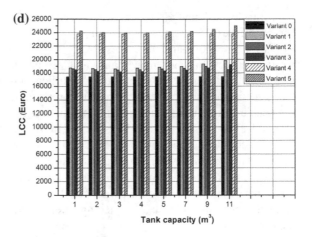

in this price was significant (Lisbon). In the case of the other four locations, i.e., Bratislava, Budapest, Warsaw, and Stockholm, the implementation of rainwater harvesting systems and graywater recycling systems was unprofitable because in these cities, unit prices and their annual increase were at a lower level, which resulted in lower annual operating costs and as a result, higher LCC values.

6.4.2 The Impact of Life Span

In order to examine the impact of the analysis period on the value of LCC costs, there were carried out the studies in which the T parameter changed. It was assumed that T will be 20 years and 50 years. Accepting a longer period of operation of the system for research, as did by other authors (Roebuck et al. 2011; Morales-Pinzón et al. 2012; Silva et al. 2015), may lead to more favorable results from the investor's point of view. This is also supported by the durability of materials currently used for building installations, mainly plastics, for which manufacturers declare a minimum service life of 50 years. For this reason, the research was conducted for the adopted initial input data which checked how the values of financial ratios characterizing the undertaking would change when the lifetime was extended to 50 years. The impact of a shorter analysis period on LCC costs was also examined, which may be significant especially in the case of variants with alternative water sources for which the LCC index value slightly differed from its level for a traditional installation solution.

The research results obtained in this area have shown that for locations where the use of alternative water system solutions was profitable already for the analysis period of 30 years, extending this period will only increase the value of the LCC indicator without changing the profitability hierarchy of these variants. It was assumed that a longer analysis period of 50 years could increase the profitability of implementing rainwater harvesting and graywater recycling systems in cities where it was unprofitable until now. However, the test results, which, for selected parameters, are shown in Fig. 6.15, did not confirm this.

As for T = 30 years, the largest LCC costs were obtained for variants with a wastewater recycling system (Variant 4 and Variant 5). Among the variants analyzed, the variant with the use of rainwater for toilets flushing, washing, and watering the garden (Variant 3) is most favorable compared to Variant 0. This is especially noticeable for four users of the installation. Such unfavorable results for variants with alternative water sources in buildings located in Budapest, Bratislava, Stockholm, and Warsaw, despite the extension of the LCC analysis period, are affected by relatively low purchase prices of water from the water supply network in these cities. In addition, the savings obtained, even over a long period of 50 years, do not compensate for the investment outlays that should be allocated to the implementation of RWHS and GWRS as well as the costs of replacing filters and pumps during the operation of these systems.

The longer analysis period had a positive effect on the results obtained for the investment located in Prague. If the building is inhabited by two people, the variant with the lowest LCC costs, as for T = 30 years, is Variant 4 (Fig. 6.16a). However, it is noticeable that the profitability of option 5 increases, in which gray wastewater is also used in addition to rainwater. The LCC costs for this solution are only 2–2.5% higher than in Variant 3, and Variant 5 can significantly reduce water consumption from the water supply network, so it is more beneficial for the environment. This variant proved to be the most cost-effective solution for this location for the case in which 4 people use the installation (Fig. 6.16b). There is a significant difference between the traditional installation and the installation additionally equipped with GWRS and RWHS (Variant 5). The LCC costs, depending on the tank capacity, are lower from 14,000 EUR to 21,000 EUR from the costs in Variant 0.

Shortening the analysis period to 20 years, in turn, resulted in the reduction of the profitability of using individual installations with alternative water sources in the case of Prague. In a situation where the installation is used by two residents, the costs of LCC Variant 0 and Variant 3 were equalized, which for T = 30 years was the most cost-effective solution (Fig. 6.17a). With a higher water consumption for four people, the LCC index value of the variants with the graywater recycling system and rainwater harvesting system increases above these costs for a traditional installation solution. The only option with lower total costs remained the option using rainwater for flushing toilets, washing, and watering the garden (Variant 3), and the optimal tank capacity is, as for T = 30 years, 7 m^3 (Fig. 6.17b).

In the case of Rome, for a 30-year analysis period, regardless of the number of people, Variant 4 and Variant 5 were unprofitable solutions. A similar trend was also maintained for a shorter time T (Fig. 6.18). Referring to the other alternative variants, it was found that if two people used the installation, they were also unprofitable, and for four users, they had lower LCC costs than Variant 0, but the cost differences are insignificant. The impact of the roof surface and the resulting charges for the discharge of rainwater into the sewage system did not have a significant impact on the results of the research in this regard.

Fig. 6.15 Results of the
financial analysis for T =
50 years for the location of
the building in **a** Budapest
(two persons), **b** Bratislava
(two persons), **c** Warsaw
(three persons), **d** Stockholm
(four persons)

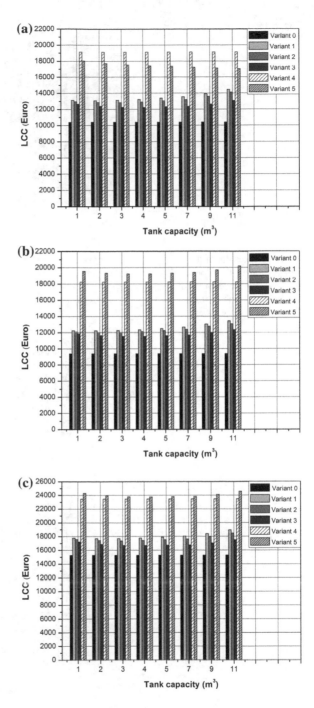

Fig. 6.15 (continued)

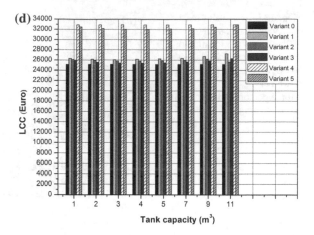

Fig. 6.16 Results of the financial analysis for T = 20 years for the location of the building in Prague, **a** two persons, **b** four persons

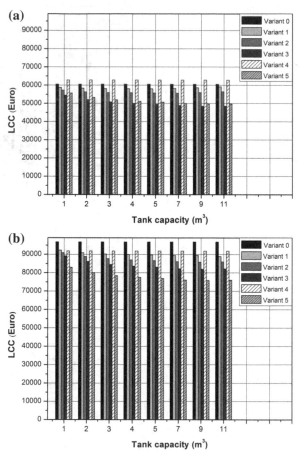

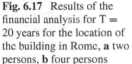

Fig. 6.17 Results of the financial analysis for T = 20 years for the location of the building in Rome, **a** two persons, **b** four persons

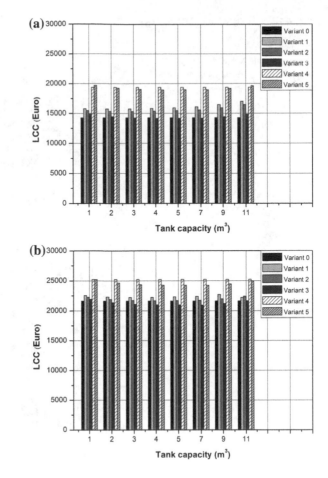

The biggest differences to the disadvantage of unconventional water installations with a shortened analysis period were observed for their location in Madrid and Lisbon. In the case of Madrid and two people using the installation, only Variant 2 and Variant 3 had lower LCC costs than Variant 0, but the differences in these costs compared to Variant 0 were only about 40 EUR and 380 EUR, for Variant 2 and Variant 3, respectively (Fig. 6.19a). As for T = 30 years, the optimal tank capacity was 7 m^3. The same capacity of the tank turned out to be the most advantageous also when four people lived in the house. Also for this number of inhabitants, the most cost-effective installation solution was the use of RWHS for flushing toilet, washing, and watering (Fig. 6.19b). In this case, with a shorter T period, there was a change in the profitability hierarchy of variants, to the disadvantage of Variant 5, which for the longer analysis period was the solution with the lowest LCC costs. This was due to increased investment outlays in Variant 5, mainly related to the implementation of the graywater recycling system, which were not compensated by the savings resulting from the reduction of water consumption from the water supply network.

Fig. 6.18 Results of the financial analysis for T = 20 years for the location of the building in Madrid, **a** two persons, **b** four persons

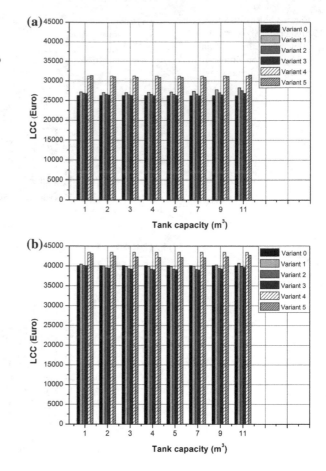

Of all the locations analyzed, reducing the T time to 20 years had the greatest impact on the results of the study obtained for Lisbon. This was due to the highest annual operating costs. For a longer analysis period, all alternative installation solutions were characterized by a lower level of LCC costs than the traditional installation variant, while for T = 20 years, the differences in the costs of individual variants were very small (Fig. 6.20). An unfavorable change is especially noticeable in the case when the building was inhabited by four people, where the difference in the LCC value between Variant 0 and the most financially profitable variant has decreased from 43,000 EUR (T = 30 years) to about 4000 EUR (T = 20 years). In addition, for the smallest number of system users, there has been a change in the profitability hierarchy of the analyzed solutions and the variants with the graywater recycling system have ceased to be financially advantageous variants. The Variant 3 with a 7 m³ tank was the most favorable in this respect. In turn, for the largest number of inhabitants, the optimal solution, as for T = 30 years, was Variant 5 with a hybrid

Fig. 6.19 Results of the financial analysis for T = 20 years for the location of the building in Lisbon **a** two persons, **b** four persons

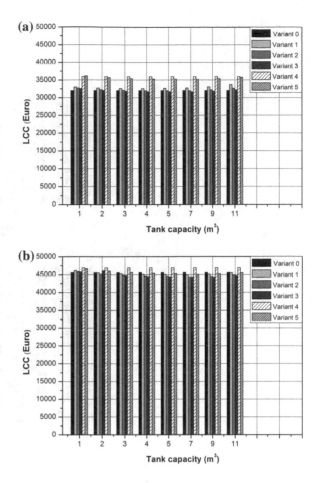

GWRS and RWHS system, but the optimal tank capacity in the rainwater harvesting system has decreased from 11 to 7 m³.

Considering that the annual increase in the purchase price of water and sewage, which was included in the study, had a significant impact on the value of total costs, a sensitivity analysis of the investment was carried out based on change, including this parameter. The results of this research are presented in Sect. 6.5.

6.5 Sensitivity Analysis

In order to assess the investment risk associated with the use of the analyzed installation variants in a single-family residential building located in various European cities, a sensitivity analysis was carried out for selected calculation cases. This analysis consisted of determining the value of total LCC costs assuming that individual

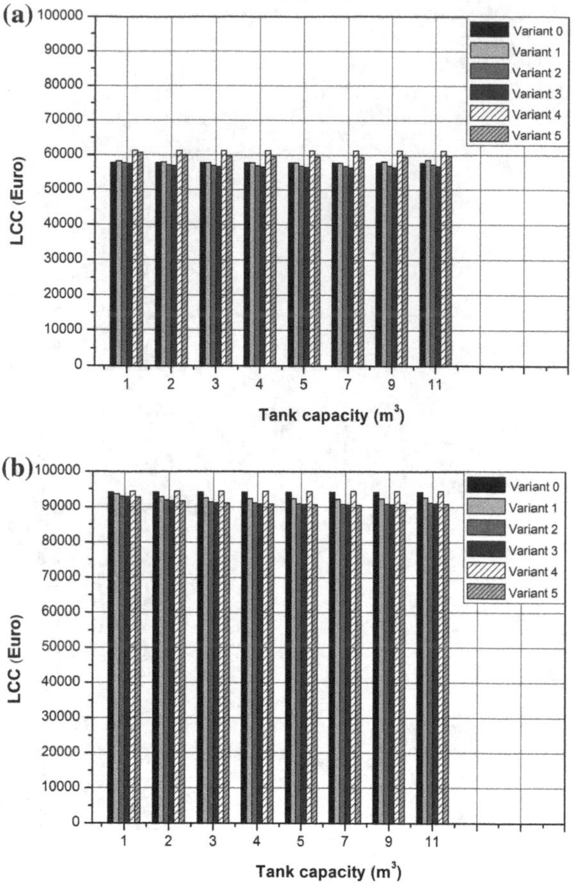

Fig. 6.20 Results of the financial analysis for T = 20 years for the location of the building in Lisbon **a** two persons, **b** four persons

components of operating costs or investments would change by a specific percentage. When analyzing the share of total OMC operating costs and INV investments in the LCC costs of the considered installation variants, it was noticed that the latter accounted for 19% to 75%, 22% to 78%, 20% to 87%, and 14% to 68%, respectively, for investments in Warsaw, Bratislava, Budapest, and Stockholm. The lower values are the INV share for Variant 0 and the maximum for Variant 5. In these locations, therefore, it was justified to examine the impact of INV changes on the LCC value, because it was this part of the total life cycle costs that determined the profitability of individual plant configurations.

In operating costs, the vast majority of them were costs related to the purchase of water and the discharge of sewage to the sewage system, while the cost of electricity associated with pumping rainwater or gray water accounted for only 2–3%. Therefore, it was assumed in the research that only the costs of purchasing water and draining sewage would change. Considering the results of the LCC analysis and the cost-effectiveness of individual installation options for each location, it was found that it would be justified to examine the impact of reducing operating costs on the total costs of LCC in Lisbon, Madrid, Rome, and Prague, i.e., in cities where the use of alternative installation solutions was cost-effective. For these four locations, investments accounted for a maximum of 20%, and in some computational cases, they were only 1%. Therefore, it was decided that a sensitivity analysis for these locations would be made taking into account the total operating costs.

The sensitivity analysis was carried out for the following ranges of changes in OMC costs or INV investments: ±10, ±25, and ±50%. The study adopted two change scenarios:

- Scenario A—change in the value of initial investments.
- Scenario B—change in the value of operating costs resulting from the amount of tap water used and the amount of sanitary wastewater discharged from the building to the sewage network.

Due to the extensive data received, this chapter presents selected results of this analysis. In the case of variants with rainwater harvesting system, the focus was on those tank volumes that proved to be optimal in the first stage of research.

The results of the research regarding changes in initial investments (Scenario A) are shown in Figs. 6.21 and 6.22. On their basis, it can be concluded that reducing the INV amount by 10% does not change the profitability hierarchy of individual variants, both for the case when the installation is used by two and four residents. The traditional option (Variant 0) was still the most cost-effective solution for all locations. A 25% reduction in investments for variants with alternative water sources

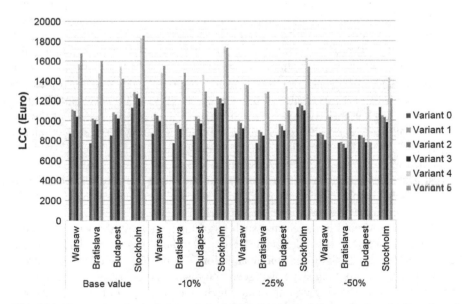

Fig. 6.21 The results of calculations of the LCC indicator assuming a decrease in the value of investments INV (Scenario A) for the case when the installation is used by two people

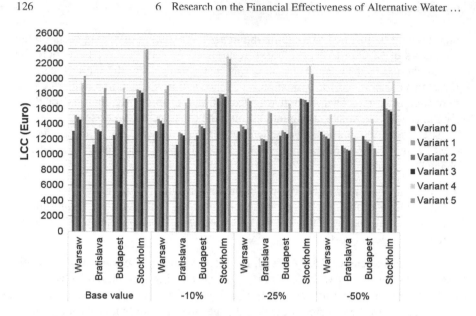

Fig. 6.22 The results of calculations of the LCC indicator assuming a reduction in the value of investments INV (Scenario A) for the case when the installation is used by four people

results in an increase in LCC costs for Variant 0, which in turn causes a change in the profitability of the investment in favor of Variant 3, but only in the case of Stockholm. For other cities, the use of unconventional installation variants in a single-family building is still financially disadvantageous. It is only by halving the investments that the financial efficiency of the analyzed options for all locations improves. In the case of Warsaw, Bratislava, and Stockholm, the most cost-effective solution is the variant, which provides for the implementation of a rainwater harvesting system for toilets flushing, washing, and watering the garden (Variant 3). However, in a situation where the investment was located in Budapest, the lowest LCC costs were obtained for two variants Variant 3 (two persons) and Variant 5 (four persons) assuming the use of rainwater and gray water. Generally, it can be stated that, with one exception, the highest LCC costs for all locations and for all computational cases were the variants equipped with a graywater recycling system. This means that due to the very high investments that are necessary to implement these systems, it is not cost-effective to use gray water as an alternative source of water in locations where there are relatively low fees for purchasing water from the network.

However, it should be noted that such large fluctuations in the value of initial investments will not take place due to the fact that their amount, in contrast to the costs associated with the use of the installation, can be determined very precisely at the investment planning stage. The sensitivity analysis carried out in this regard, however, allowed drawing an important conclusion, namely cofinancing for costs allocated to the implementation of alternative water systems may increase their financial efficiency and contribute to their wider use in countries where they are rarely

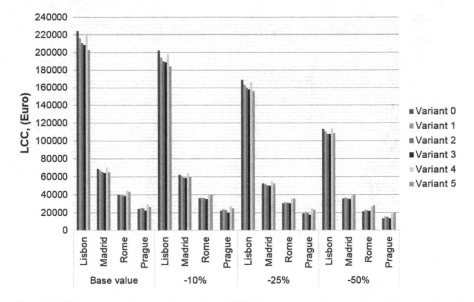

Fig. 6.23 The results of calculations of the LCC indicator assuming a reduction in the value of operating costs (Scenario B) for the case when the installation is used by two people

used. Such funding as an element of pro-ecological policy has been practiced for many years in various countries around the world.

In the case of cities where investments constituted an insignificant part of LCC costs, their change would not cause significant differences in the value of total life cycle costs, and thus would not affect the profitability hierarchy of individual installation variants. Taking this into account, research on investment sensitivity for these locations was carried out, taking into account the reduction of operating costs by 10%, 25%, and 50% that could be caused in the future, less than expected increase in water purchase prices and prices for sewage disposal into the sewage system. The results of this analysis are shown in Figs. 6.23 and 6.24. They clearly show that the investment based on the implementation of systems with alternative water sources in Lisbon is not sensitive to changes according to Scenario B. Even reducing the operating costs by half will not change the most cost-effective option, i.e., Variant 5, which brings the greatest water savings. In the case of Rome, where the differences in LCC costs between the considered variants were a slight reduction in costs in the examined area, it improves the financial efficiency of the traditional installation solution (Variant 0), to the detriment of the variant with the rainwater harvesting system (Variant 3). The situation is similar in the case of Prague. Madrid was the second city after Lisbon with the largest share of operating costs in the total costs of LCC. However, the location of alternative water system variants in its area is more risky than in Lisbon, with a 50% reduction in operating costs, the LCC indicator for Variant 0 reaches the same level as for variants with RWHS and GWRS, especially for four users.

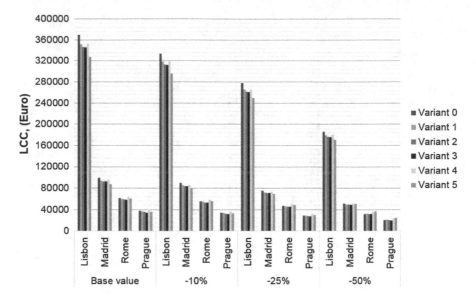

Fig. 6.24 The results of calculations of the LCC indicator assuming a reduction in the value of operating costs (Scenario B) for the case when the installation is used by four people

Variant 0, i.e., the traditional installation layout, always has the lowest investments. As a result, most investors are currently choosing such a solution for plumbing and building installations, without analyzing the operating costs associated with the operation of the system in the long run. Meanwhile, the Life Cycle Cost analysis showed that operating costs can determine the level of the LCC indicator and decide on the financially most favorable option. It is especially observed for cities with high fees for water and sewage disposal and/or their annual increase is significant. In the study, these were the following locations: Lisbon, Madrid, Rome, and Prague. It was in these cities that the implementation of alternative solutions for water installations supplied with rainwater and gray water was financially profitable. The necessity to incur additional investments for rainwater harvesting system (RWHS) and graywater recycling system (GWRS) and the lack of coverage by savings resulting from their use meant that unconventional installation systems were unprofitable in Warsaw, Bratislava, Budapest, and Stockholm.

The sensitivity analysis conducted also showed that the results obtained in the first stage could be considered correct, and the investments consisting in the implementation of RWHS and GWRS were slightly susceptible to changing individual parameters of the financial model. No significant changes in the profitability hierarchy of the installation variants were observed.

The results of the LCC analysis also have a practical aspect and can be a guide for potential investors and operators to apply this type of system already at the investment planning stage. An additional impulse to use them, especially in locations where their use is not profitable, could also be funding from the state budget or from pro-ecological funds of international organizations, as is the case in many other countries.

References

Bakis N, Kagiouglou M, Aouad G, Amaratunga D, Kishk M, Al-Hajj A (2003) An integrated environment for life cycle costing in construction

Brigham E, Ehrhardt M (2008) Financial management. Theory and PRACTICE. South-Western Cengage Learning, Ohio, USA

Buildings... (2008) Buildings and constructed assets—Service life planning. Part 5: life cycle costing. ISO/DIS 15686-5.2. International Organization for Standardization

Clift M (2003) Life-cycle costing in the construction sector. Sustainable building and construction. UNEP Industry and Environment

Dell Isola A, Kirk S (1981) Life cycle costing for design professionals. McGraw-Hill, New York

DOE (2014) Life cycle cost handbook. Guidance for life cycle cost estimation and analysis. Office of acquisition and project management, U.S. Department of Energy, Washington

Epstein M (1996) Measuring corporate environmental performance. McGraw-Hill, Chicago, IL

Flanagan R, Norman G (1987) Life cycle costing: theory and practice, RICS. Surveyors Publications Ltd., London

Flanagan R, Kendell A, Norman G, Robinson G (1987) Life cycle costing and risk management. Construction Management and Economics, No 5

Fuller S, Petersen S (1996) Life cycle costing manual for the federal energy management program. NIST, Handbook, p 135

Gluch P, Baumann H (2004) The life cycle costing (LCC) approach: a conceptual discussion of its usefulness for environmental decision-making. Build Environ 39

Góralczyk M, Kulczycka J (2003) LCNPV as a tool for evaluation of environment al investment in industrial Project. In: 2nd international symposium ILCDES 2003—Integrated life-time engineering of buildings and civil infrastructures, Kuopio

Haviland D (1978) Life cycle cost analysis. Using it in practice. AIA

Kowalski Z, Kulczycka J, Góralczyk M (2007) Ekologiczna ocena cyklu życia procesów wytwórczych. Wydawnictwo Naukowe PWN, Warszawa

Landres R (1996) Product assurance dictionary. Marlton Publishers, New York

Lang A (2007) Sensitivity analysis of life cycle cost calculation. http://www.ee.kth.se/php/modules/publications/reports/2007/IR-EE-ETK_2007_010.pdf

Liaw C, Tsai Y (2004) Optimum storage volume of rooftop rain water harvesting systems for domestic use. J Am Water Resour Assoc 40:901–912

Life Cycle Cost (LCC) (2011) Description of the tool and its parameters. Swedish Environmental Management Council

Life cycle costing... (2007a) Life cycle costing (LCC) as a contribution to sustainable construction: a common methodology, Literature review. http://ec.europa.eu/enterprise/sectors/construction/files/compet/life_cycle_costing/literat_review_en.pdf

Life Cycle Costing... (2007b) Life cycle costing (LCC) as a contribution to sustainable construction: a common methodology. http://ec.europa.eu/enterprise/sectors/construction/files/compet/life_cycle_costing/common_methodology_en.pdf

Life cycle costing... (2007c) Life cycle costing (LCC) as a contribution to sustainable construction, Guidance on the use of the LCC Methodology and its application in public procurement http://ec.europa.eu/enterprise/sectors/construction/files/compet/life_cycle_costing/guidance__case_study_en.pdf

Military Handbook... (1983a) Military handbook, life cycle cost in navy acquisitions, MIL-HDBK-259

Military Handbook... (1984a) Military handbook, life cycle cost model for defense material systems, MIL-HDBK-276-1

Military Handbook... (1984b) Military handbook, life cycle cost model for defense material systems operating instructions, MIL-HDBK-276-2

Morales-Pinzón T, Lurueña R, Rieradevall J, Gasol CM, Gabarrell X (2012) Financial feasibility and environmental analysis of potential rainwater harvesting systems: a case study in Spain. Resour Conserv Recycl 69:130–140

Pastusiak R (2003) Ocena efektywności inwestycji. CeDewu Sp. zoo, Warszawa

Rahman A, Dbais J, Imteaz M (2010) Sustainability of rainwater harvesting systems in multistorey residential buildings. Am J Appl Sci 3:889–898

Rahman A, Keane J, Imteaz AM (2012) Rainwater harvesting in greater sydney: water savings, reliability and economic benefits. Resour Conserv Recycl 61:16–21

Roebuck RM, Oltean-Dumbrava C, Tait S (2011) Whole life cost performance of domestic rainwater harvesting systems in the United Kingdom. Water Environ J 25:355–365

Rogowski W (2004) Rachunek efektywności inwestycji. Oficyna Ekonomiczna, Kraków

SAE (1999) Reliability and maintainability guideline for manufacturing machinery and equipment. Society of Automotive Engineers (SAE), M-110.2

Silva C, Sousa V, Carvalho N (2015) Evaluation of rainwater harvesting in Portugal: application to single-family residences. Resour Conserv Recycl 94:21–34

Task Group 4 (2003) Life cycle costs in construction. Final Report. http://www.ceetb.eu/docs/Reports/LCC%20FINAL%20REPORT-2.pdf

White G, Ostwald P (1976) Life cycle costing. Management Accounting

Woodward D, Demirag I (1989) Life cycle costing. Career Accountant

Chapter 7
The Impact of Rainwater Harvesting on a Drainage System and a Catchment

Abstract The use of RWHS not only protects natural water resources, brings financial benefits resulting from the lower consumption of tap water, but also by reducing the outflow of water from the catchment, it also has a positive effect on the functioning of sewerage systems. Taking the above into account, the research was conducted to determine the impact of RWHS application on the volume of rainwater outflow from the catchment and on the functioning of the sewage system. A real urban catchment was selected for the study, for which a hydrodynamic model was developed in the stormwater management Model (SWMM).

Transformation of natural areas into residential and industrial ones causes large environmental and hydrological changes (Gunn et al. 2012; O'Driscoll et al. 2010). These changes have a negative impact primarily on the quantity and quality of rainwater causing an increase in the speed and volume of runoff, a reduction in infiltration, an increased risk of flooding, and hydraulic overloading of sewer systems (Stec and Słyś 2013). In order to limit these negative changes, some actions are taken that consist of applying the principle of sustainable development in the aspect of spatial planning and rainwater management. These include primarily objects and devices that reduce and delay the time of outflow of water from the catchment through their infiltration and retention (Pochwat et al. 2017; Pochwat 2017; Starzec et al. 2018; Starzec and Dziopak 2018).

In recent years, there have been many strategies for rainwater management in urban areas, which, unlike traditional drainage systems, are more environment-friendly. One of them is the low impact development (LID) strategy, which, through the use of appropriate techniques, aims at restoring the hydrological state of a catchment before its management (USEPA 2000). Low-impact development practices mainly include decentralized devices and objects whose operation is to imitate the natural hydrological processes taking place in the catchment such as infiltration, evaporation, and retention of rainwater. One of the LID practices is rainwater harvesting systems (RWHS) (Palla et al. 2017).

The use of RWHS not only protects natural water resources but also brings financial benefits resulting from the lower consumption of tap water, which was discussed in the previous chapters. By reducing the outflow of water from the catchment, it also

© Springer Nature Switzerland AG 2020

A. Stec, *Sustainable Water Management in Buildings*,
Water Science and Technology Library 90,
https://doi.org/10.1007/978-3-030-35959-1_7

has a positive effect on the functioning of sewerage systems and objects interacting with them (Freni and Liuzzo 2019; Teston et al. 2018). Various studies have shown that in urban catchments, the RWHS installation could be an effective support for reducing the frequency and peak of stormwater flood (Gerolin et al. 2010; Zhang et al. 2012). Teston et al. (2018) examined the impact of rainwater harvesting systems on the sewage flow of sewerage and found that there was a reduction in peak flow, but its size depended on the capacity of the tank, rainfall characteristics, and parameters related to the building.

Taking the above into account, the research was conducted to determine the impact of RWHS application on the volume of rainwater outflow from the catchment and on the functioning of the sewage system. A real urban catchment was selected for the study, for which a hydrodynamic model was developed in the stormwater management Model (SWMM). The research did not take into account the graywater recycling systems since it is mainly rainwater in the rapidly changing urban catchments that are perceived as a big problem that needs to be solved. Negative changes in urban areas and the associated increase in stormwater runoff, not only affect the natural environment, but through increasingly frequent urban floods and outflows from the sewage system, it generates very large material losses.

7.1 Study Area

A municipal drainage catchment located in the southeastern part of Poland was selected for the research. It is a city of Przemyśl in both sides of the San River which belongs to the European Ecological Natura 2000 Network and is the main receiver of sewage from the city (Fig. 7.1). Its area covers 46 km^2 and is inhabited by over 65,000 residents. Due to the high degree of urbanization, the left-bank part of the city covering the district of Zasanie was selected for the research. There is a combined sewerage system in its area, which in rainy weather conditions, affects significantly the operation of wastewater treatment plants and the quality of water in the San River. The main problem in the operation of this sewage system is the occurrence of pressure flows in the main sewage collectors and an excessive number of discharges of storm sewage into the river. Based on the maps and design of the existing sewerage system, a hydrodynamic model of the analyzed catchment was developed. The district of Zasanie model developed in the stormwater management model (SWMM) is shown in Fig. 7.2. The total area of this part of the city is 632,88 ha. It mainly consists of residential and service areas.

When analyzing the area of the district of Zasanie using an orthophotomap, an area of over 16 ha was identified on two of the subdivisions (S1 and S2), where 366 single-family residential lots were identified (Fig. 7.2). Then, on these plots, the surface of rooftops, other sealed areas, and green areas were measured. The area of the plots ranged from 362 to 1700 m^2 (632 m^2 on average) and the rooftops area from 64 to 240 m^2 (132 m^2 on average). The detailed data characterizing the analyzed part

Fig. 7.1 Location of the case study city in Poland

of the catchment, which were introduced to the SWMM program, are presented in Table 7.1.

In order to take into account the uncertainty associated with the residents' acceptance for installing RWHS in their homes, various scenarios for their implementation in the catchment analyzed were adopted, which are presented in Table 7.2. Scenario 0 means no rainwater harvesting systems are used, which corresponds to the current state. The division into such scenarios will enable a detailed examination of the impact of the economic use of rainwater in the analyzed catchment on the sewage system.

The research took into account, the test results described in Chaps. 5 and 6 regarding the capacity of the tanks and the related rainwater harvesting systems efficiency. Local conditions resulting from the location of buildings in the catchment analyzed were also taken into account. Considering also that the use of rainwater as an alternative source of water in Polish conditions is not financially viable, as demonstrated by the research, in the hydrodynamic model, on the average, from among the analyzed capacities, the reservoir capacity of 5 m^3 was assumed.

Table 7.1 Basic input data
for hydrodynamic model of
the catchment analyzed

Parameter	Value
Land surface slope	0.5–5.0%
Manning's coefficient for impervious surfaces	0.015
Manning's coefficient for pervious surfaces (e.g. dense grass)	0.25
Impervious depression storage	1.5 mm
Pervious depression storage	7.0 mm
Percent imperviousness: Rooftop Green areas Other impervious surfaces	100% 10–20% 40–70%

Table 7.2 Scenarios for RWHS implementation

Scenario	Implementation, %	Households
Scenario 0	0	0
Scenario 1	50	183
Scenario 2	100	366

7.2   Results of Hydrodynamic Simulations

Simulation research for real rainfall data from 2007 to 2008 showed that the use of rainwater harvesting systems in the analyzed catchment would reduce the outflow of rainwater to the sewage system and increase the infiltration of these waters to the ground.

When analyzing the change in hydrological conditions in the catchment depending on the number of RWHS applied (Scenario 1 and Scenario 2), it was noted that the peak runoff was reduced in the range from 21% to even 100% and from 25% also to 100%, respectively, for scenario 1 and 2 in relation to scenario 0 (the catchment in the existing state without RWHS). The sample results in this area are shown in Figs. 7.3 and 7.4. The total reduction of the peak runoff was observed for negligible precipitation with an intensity of no more than 2 mm/h. Precipitation with an intensity of 2–10 mm/h caused an increase in the amount of sewage flowing from the catchment and a reduction of this outflow at the level of about 28%. In turn, in the case of rainfall of more than 10 mm/h, the outflow of rainwater was reduced by about 26% on average. The results were also affected by the length of the rain-free period between successive rainfalls and the possibility of rainwater holding in reservoirs in rainwater harvesting systems.

The assumption considered in the model that underdrain outflow and overflow from the RWHS will be directed to the permeable area (green areas around the

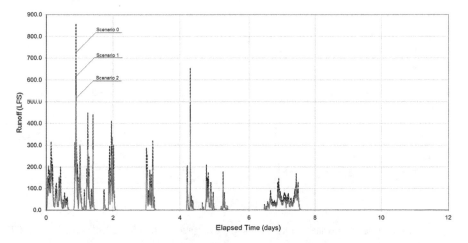

Fig. 7.3 Rainwater peak runoff from S1 for precipitation from September 5–12, 2007

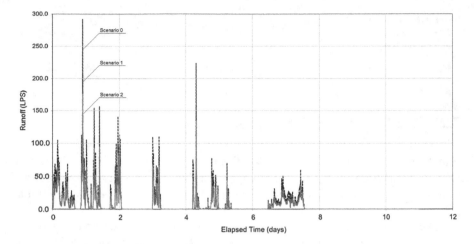

Fig. 7.4 Peak runoff from S2 for precipitation from September 5–12, 2007

buildings) caused an increase in the volume of rainwater which they infiltrated into the ground. If in the catchment, RWHS in 50% households (Scenario 1) is implemented, then infiltration will increase from 9% even in some cases up to 400% compared to Scenario 0. The average increase for this scenario was about 30%, while for scenario 2 (100% RWHS implementation), an average of 35%. Selected test results showing the infiltration process depending on the duration of rainfall are shown in Fig. 7.5.

Considering the fact that in the sewage system analyzed, there are periodically pressure sewage flows, one of the research objectives was to determine the impact of rainwater harvesting systems on the hydraulic conditions of this network. However,

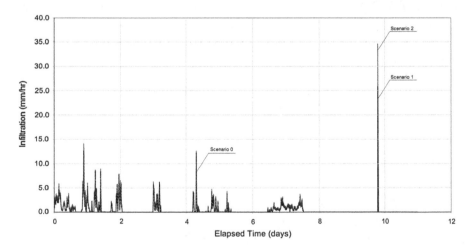

Fig. 7.5 Infiltration of rainwater in the catchment S2 for precipitation from September 5–12, 2007

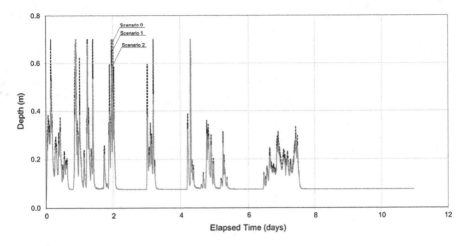

Fig. 7.6 Filling in the main canal located below the catchment analyzed for precipitation from September 5–12, 2007

the tests carried out on the hydrodynamic model showed that for the RWHS parameters adopted, their influence on the reduction of hydraulic overload of the canals was insignificant. The implementation of RWHS in even all households (Scenario 2) located in the catchment area under consideration will not eliminate the phenomenon of pressure flows in the main canals, and will only shorten the time of their occurrence (Fig. 7.6).

The reduction of rainwater flow in the drainage network located below the considered catchment ranged from 1 to 25%, on average, 6% for Scenario 1 and from 1 to 39%, on average, 8% for Scenario 2. Some sample results in this area are shown in Fig. 7.7.

There were no significant differences in the functioning of the sewage system on further sections of the sewerage network, especially those located in front of the main sewage pumping station and in the reduction of discharges via stormwater transfers to the San River. Main canals supplying sewage to the pumping stations very often in the current state are hydraulically overloaded and there are periodically pressure sewage flows in them. This is due to the fact that the sewerage system, especially in the central part of the city, was designed for conditions existing several decades ago and in the current phase of the city's development, it is not able to accept all of the falling rainwater. The results of hydrodynamic simulations showed that the implementation of RWHS only in a small area of the catchment would not reduce significantly the occurrence of these adverse phenomena at the critical points of the sewer system under investigation. Figure 7.8 shows the level of sewage filling in one of the main canals, where the moments in which the canal operates under pressure are visible.

The results of the research conducted showed that the use of rainwater harvesting systems (RWHS) to capture rainwater discharged from the roof of the building was not effective in reducing the occurrence of pressure flows in the sewage system

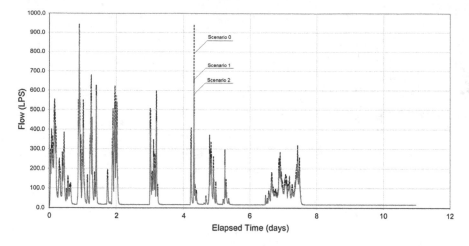

Fig. 7.7 Sewage flow in the main canal located below the catchment analyzed for precipitation from September 5–12, 2007

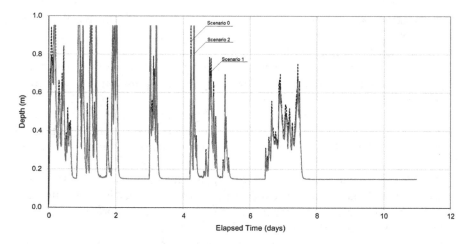

Fig. 7.8 Filling in the main canal located before the pumping station for precipitation from September 5–12, 2007

analyzed. Many factors could have influenced such results, including too small numbers of RWHS in relation to the total area of the Zasanie catchment. Such systems should be included in the whole catchment, in buildings where it would be possible to implement them or additionally adopt other LID solutions that increase retention and infiltration of rainwater.

However, it should be noted that due to the RWHS implementation in the catchment analyzed, there was a reduction in the outflow of water, their retention in the place where the precipitation occurred, and then the intensification of their infiltration into the ground. This is a very beneficial phenomenon in improving the hydrological conditions in the basin and contributing to the protection of water resources. Therefore, rainwater harvesting systems meet the basic assumptions of sustainable rainwater management in urbanized areas.

References

Freni G, Liuzzo L (2019) Effectiveness of rainwater harvesting systems for flood reduction in residential urban areas. Water 11(7):1389

Gerolin A, Kellagher RB, Faram MG (2010) Rainwater harvesting systems for stormwater management: feasibility and sizing considerations for the UK. In: Proceedings of the Novatech 2010—7th international conference on sustainable techniques and strategies for urban water management, Lyon, France

Gunn R, Martin A, Engel B, Ahiablame L (2012) Development of two indices for determining hydrologic implications of land use changes in urban areas. Urban Water J 9(4):239–248

O'Driscoll M, Clinton S, Jefferson A, Manda A, McMillan S (2010) Urbanization effects on watershed hydrology and in-stream processes in the southern United States. Water 2(3):605–648

Palla A, Gneco I, La Barbera P (2017) The impact of domestic rainwater harvesting systems in storm water runoff mitigation at the urban block scale. J Environ Manag 191:297–305

Pochwat K (2017) Hydraulic analysis of functioning of the drainage channel with increased retention capacity. In: E3S web of conferences 17, 00075. https://doi.org/10.1051/e3sconf/20171700075

Pochwat K, Słyś D, Kordana S (2017) The temporal variability of a rainfall synthetic hyetograph for the dimensioning of stormwater retention tanks in small urban catchments. J Hydrol 549:501–511

Starzec M, Dziopak J (2018) Designing a retention sewage canal with consideration of the dynamic movement of precipitation over the selected urban catchment. Undergr Infrast Urban Areas 4:181–191

Starzec M, Dziopak J, Słyś D (2018) The impact of the channel retention before the tank on its retention capacity. Undergr Infrastruct Urban Areas 4:193–200

Stec A, Słyś D (2013) Effect of development of the town of Przemysl on operation of its sewerage system. Ecol Chem Eng S 20:381–396

Teston A, Teixeira CA, Ghisi E, Cardoso EB (2018) Impact of rainwater harvesting on the drainage system: case study of a condominium of houses in Curitiba. Southern Brazil Water 10:1100

USEPA (US Environmental Protection Agency) Low Impact Development (LID) (2000) A literature review office of water. Washington, D.C, EPA-841-B-00-005

Zhang X, Hu M, Chen G, Xu Y (2012) Urban rainwater utilization and its role in mitigating urban waterlogging problems—A case study in Nanjing. China Water Resour Manag 26:3757–3766

Chapter 8
Awareness and Acceptance of the Public Regarding Alternative Water Sources

Abstract In recent years, there have been publications presenting the opinions of various communities regarding their adoption and implementation of centralized and decentralized rainwater harvesting and graywater recycling systems, but this was mainly for countries outside Europe. Considering the above, the research was carried out in selected European countries whose aim was to determine the factors influencing society's reluctance to use gray water and rainwater. The analysis of the results allowed determining the level of awareness about water shortages, water-saving behavior, and the attitude of respondents to alternative water sources. The analysis also took into account demographic and external factors such as geographical location and climate, which may affect the results.

Technologies of alternative water sources used to supply buildings with water are currently being researched, improved, and implemented in many countries around the world. The main purpose of the research is to determine whether alternative water supply systems are technically feasible, economically justified, and also socially acceptable. Researchers have devoted a lot of attention to technical and economic issues, while issues of public awareness and its acceptance regarding the use of alternative water sources are still often overlooked. In recent years, there have been publications presenting the opinions of various communities regarding their adoption and implementation of centralized and decentralized rainwater harvesting and graywater recycling systems, but this was mainly for countries outside Europe (Garcia-Cuerva et al. 2016; Gu et al. 2015; Dolnicar and Schäfer 2009; Skoien and Gardner 2008). These studies show that there is an important sociocultural element that can influence people's decisions to use alternative water sources. Some researchers note that the key in this respect is low awareness of people resulting from the lack of appropriate information campaigns (Dolnicar et al. 2010). This is also confirmed by cases where public opposition has led to delay or failure to implement systems powered by unconventional water sources (Fielding et al. 2018; Hurlimann and Dolnicar 2010).

Considering the above, the research was carried out in selected European countries whose aim was to determine the factors influencing society's reluctance to use recovered gray water and rainwater, despite the fact that in many regions of Europe

A. Stec, *Sustainable Water Management in Buildings*,
Water Science and Technology Library 90,
https://doi.org/10.1007/978-3-030-35959-1_8

potable water resources are very poor. The analysis of the results allowed determining the level of awareness about water shortages, water-saving behavior, and the attitude of respondents to alternative water sources. The analysis also took into account demographic and external factors such as geographical location and climate, which may affect the results.

8.1   A Description of Research Methodology

In order to conduct the research, a questionnaire was prepared consisting of 13 questions, 4 regarding general information about the people surveyed and 9 related to the subject of water saving and the possibilities of using alternative sources. The survey was conducted in 485 households located in selected European countries. Due to the subject of the survey, statistical units for the sample were selected in a targeted manner. Targeted selection is based on the researcher's conscious decision about the selection of population elements.

The statistical package for social sciences (SPSS) program was used to develop the survey results, which is a commonly used tool for conducting statistical analyses in both social and technical sciences (Ong and Puteh 2017; Arrkelin 2014).

Due to the fact that the results of the survey allowed determining mainly statistical nonmeasurable (qualitative) features, the significance of differences between qualitative (nominal) variables was checked using the Chi-square independence test. The formula (8.1) was used to calculate its value. In statistical analyses, a significance level of $p < 0.05$ was adopted, similar to the studies presented in (Garcia-Cuerva et al. 2016).

$$\chi^2 = \sum_{i=1}^{n} \frac{(O_i - E_i)^2}{E_i} \tag{8.1}$$

where O_i—observed value, E_i—expected value.

8.2   Survey Results and Their Analysis

8.2.1   Research Group Analysis

The survey involved 485 respondents, the majority of respondents came from Poland (14%) and the Czech Republic (13.4%), the least from countries such as Portugal (11%) and Hungary (11%) (Table 8.1).

In the group analyzed, men slightly dominated (51%), the percentage of women was less than 49%. A similar distribution was recorded in most countries, except for the Czech Republic and Italy, with slightly more female respondents from these

Table 8.1 Share of respondents in individual countries

Country	Frequency	Percentage
The Czech Rep.	65	13.4
Spain	60	12.4
Poland	70	14.4
Portugal	55	11.3
Slovakia	70	14.4
Sweden	50	10.3
Hungary	55	11.3
Italy	60	12.4
Total	485	100.0

countries. Statistical analysis using the independence test showed no statistically significant differences between countries in terms of gender ($p = 0.973$) (Table 8.2).

Table 8.2 Gender of respondents divided into the country of residence

			Gender of respondents	
			Female	Male
Country	The Czech Republic	Population	33	32
		%	50.8	49.2
	Spain	Population	29	31
		%	48.3	51.7
	Poland	Population	34	36
		%	48.6	51.4
	Portugal	Population	27	28
		%	49.1	50.9
	Slovakia	Population	34	36
		%	48.6	51.4
	Sweden	Population	22	28
		%	44.0	56.0
	Hungary	Population	25	30
		%	45.5	54.5
	Italy	Population	33	27
		%	55.0	45.0
Total		Population	237	248
		%	48.9	51.1
Chi-square independence test			$\chi^2 = 1.74; p = 0.973$	

The majority of people were up to 35 years old (41%) and between 35 and 45 years old (37%), the percentage of people over 45 years was slightly over 22%. A similar distribution was recorded among respondents from Poland, Portugal, and Italy. Among the respondents from the Czech Republic, the percentage of people aged up to 35 and aged 35–45 was the same. In turn, the respondents from Spain, Slovakia, and Hungary were dominated by people aged 35–45. Statistical analysis using the independence test showed no statistically significant differences between countries in terms of age ($p = 0.776$) (Table 8.3).

Among the respondents, people with higher education predominated, both when all the respondents were considered and broken down by the country. Statistical analysis using the independence test did not show statistically significant differences between countries in terms of education ($p = 0.793$) (Table 8.4). Taking into account the respondent's place of residence, people living in cities predominated. Both when all the respondents were considered and by country. Statistical analysis using the independence test also did not show statistically significant differences ($p = 0.500$) in this case (Table 8.5).

Table 8.3 Age of respondents divided into the country of residence

			Age of respondents		
			<35 yrs old	From 35 to 45 yrs old	≥45 yrs old
Country	The Czech Republic	Population	25	25	14
		%	39.1	39.1	21.9
	Spain	Population	20	27	13
		%	33.3	45.0	21.7
	Poland	Population	33	27	10
		%	47.1	38.6	14.3
	Portugal	Population	23	18	14
		%	41.8	32.7	25.5
	Slovakia	Population	27	24	18
		%	39.1	34.8	26.1
	Sweden	Population	19	20	11
		%	38.0	40.0	22.0
	Hungary	Population	22	23	10
		%	40.0	41.8	18.2
	Italy	Population	27	16	17
		%	45.0	26.7	28.3
Total		Population	196	180	107
		%	40.6	37.3	22.2
Chi-square independence test			$\chi^2 = 9.82; p = 0.776$		

Table 8.4 Education of respondents divided into the country of residence

			Education of respondents	
			Secondary	Higher
Country	The Czech Republic	Population	19	44
		%	30.2	69.8
	Spain	Population	16	44
		%	26.7	73.3
	Poland	Population	19	51
		%	27.1	72.9
	Portugal	Population	14	41
		%	25.5	74.5
	Slovakia	Population	15	55
		%	21.4	78.6
	Sweden	Population	8	42
		%	16.0	84.0
	Hungary	Population	14	41
		%	25.5	74.5
	Italy	Population	16	44
		%	26.7	73.3
Total		Population	121	362
		%	25.1	74.9
Chi-square independence test			$\chi^2 = 3.87; p = 0.793$	

To sum up, the analyzed groups from individual countries were comparable in terms of sex, age, education, and place of residence.

8.2.2 Possibilities of Using Alternative Sources in a Building

The second part of the survey, comprising 9 questions, allowed determining the level of knowledge of the respondents about the abundance of their countries in natural water resources and the possibilities of their protection through the use of alternative water sources.

Slightly more than half of the respondents thought that there was a shortage of drinking water in their country (50%). Most often the inhabitants of Spain (76%) and Portugal (76%) thought so. The smallest percentage of people who think that there is a shortage of drinking water in their country came from Sweden (16%) and Slovakia (37%). The difference in responses between respondents from individual countries is statistically significant ($p < 0.001$) (Table 8.6).

Table 8.5 Place of residence of respondents divided into the country of residence

			Place of residence	
			Country	City
Country	The Czech Republic	Population	30	35
		%	46.2	53.8
	Spain	Population	21	39
		%	35.0	65.0
	Poland	Population	27	43
		%	38.6	61.4
	Portugal	Population	23	32
		%	41.8	58.2
	Slovakia	Population	19	51
		%	27.1	72.9
	Sweden	Population	17	33
		%	34.0	66.0
	Hungary	Population	22	33
		%	40.0	60.0
	Italy	Population	22	38
		%	36.7	63.3
Total		Population	181	304
		%	37.3	62.7
Chi-square independence test			$\chi^2 = 6.34; p = 0.500$	

The research also analyzed the impact of countries' water resources on the responses of their inhabitants. It was divided into two groups: countries with resources less than 10,000 m^3/person and countries where water resources exceed 10,000 m^3/person. The first group includes Poland, the Czech Republic, Italy, Portugal, and Spain. On the other hand, Slovakia, Sweden, and Hungary. Chi-square analysis showed a statistically significant difference between respondents from countries with lower water resources per capita (<10,000 m^3), and respondents from countries where water resources per capita exceed 10,000 m^3 ($p < 0.001$). Water shortage was most noticed by people from countries with low resources (57%) than people from countries with large resources of potable water (38%) (Table 8.7).

Respondents' responses were also assessed regarding the impact of the climate in which the country is located. In the temperate climate there are countries such as the Czech Republic, Poland, Slovakia, Sweden, and Hungary. The climate of Spain, Portugal, and Italy was defined as Mediterranean. Chi-square analysis showed a statistically significant difference between respondents from countries with a moderate climate and respondents from countries with a Mediterranean climate ($p < 0.001$). Water shortage was most noticed by people from Mediterranean countries (67%) than people from countries with moderate climate (41%) (Table 8.8).

Table 8.6 Does the respondent think that in their country there is a problem of potable water shortage; a division into countries

			Problem of potable water shortage	
			Yes	No
Country	The Czech Republic	Population	28	37
		%	43.1	56.9
	Spain	Population	46	14
		%	76.7	23.3
	Poland	Population	32	38
		%	45.7	54.3
	Portugal	Population	42	13
		%	76.4	23.6
	Slovakia	Population	26	44
		%	37.1	62.9
	Sweden	Population	8	42
		%	16.0	84.0
	Hungary	Population	32	23
		%	58.2	41.8
	Italy	Population	29	31
		%	48.3	51.7
Total		Population	243	242
		%	50.1	49.9
Chi-square independence test			$\chi^2 = 63.40; p < 0.001$	

Table 8.7 Does the respondent think that in their country there is a problem of potable water shortage; a division according to the country's water resources

			Response	
			Yes	No
Country water resources (calculated per one inhabitant)	<10 000 m^3	Population	177	133
		%	57.1	42.9
	>10 000 m^3	Population	66	109
		%	37.7	62.3
Chi-square independence test			$\chi^2 = 16.81; p < 0.001$	

Table 8.8 Does the respondent think that in their country there is a problem of potable water shortage; a division according to the country's climate

			Response	
			Yes	No
Country climate	Moderate	Population	126	184
		%	40.6	59.4
	Mediterranean	Population	117	58
		%	66.9	33.1
Chi-square independence test			$\chi^2 = 30.74; p < 0.001$	

In a situation where all respondents were considered, the place of residence (city or village) would not significantly differentiate the perception of water shortages ($p = 0.418$). However, such a difference was found in the group of respondents from Portugal ($p = 0.027$) and Italy ($p = 0.019$). Respondents from these countries, living in rural areas more often (respectively, 91% and 68% among them) noticed the problem of drinking water shortage than residents of cities from these countries (66% and 37%, respectively) (Table 8.9).

Table 8.9 Does the respondent think that in their country there is a problem of potable water shortage by country and place of residence; % of respondents answered "yes"

		Place of residence		Chi-square ($\chi^2; p$)
		Country	City	
The Czech Republic	Population	13	15	0.00
	% of place	43.3%	42.9%	0.969
Spain	Population	16	30	0.00
	% of place	76.2%	76.9%	0.949
Poland	Population	9	23	2.71
	% of place	33.3%	53.5%	0.099
Portugal	Population	21	21	4.89
	% of place	91.3%	65.6%	0.027
Slovakia	Population	5	21	1.31
	% of place	26.3%	41.2%	0.253
Sweden	Population	2	6	0.34
	% of place	11.8%	18.2%	0.558
Hungary	Population	14	18	0.45
	% of place	63.6%	54.5%	0.503
Italy	Population	15	14	5.48
	% of place	68.2%	36.8%	0.019
Total	Population	95	148	0.66
	% of place	52.5%	48.7%	0.418

The respondents also answered the question about the reasons for their saving of water. They could indicate a desire to reduce water charges, the need to protect water resources, or both. The most common reason for saving water was the protection of water resources (65%), slightly less often such a reason was the desire to reduce bills (54%). The most frequently wanted to protect water resources were respondents from Sweden (84%), Spain (77%), Portugal (75%), the least frequently surveyed from the Czech Republic (49%), Hungary (53%), and Italy (57%). The difference is statistically significant ($p < 0.001$). Reduction of bills—most frequently indicated by respondents from the Czech Republic (68%), Italy (67%), Hungary (64%), Poland (63%), the least often this reason was chosen by respondents from Spain (32%), Sweden (34%), and Portugal (38%). The difference between the groups is statistically significant ($p < 0.001$) (Table 8.10).

The vast majority of respondents said that their households save water (84%). This was most often indicated by respondents from Spain (92%), Sweden (88%), Portugal (87%), and the least frequently surveyed from Slovakia (76%). The difference between the groups was not significant and there were no statistically significant differences between them ($p = 0.288$) (Table 8.11).

Table 8.10 Reasons for saving water,% of indications by a country

			Water resources protection	Reduction of bills
Country	The Czech Republic	Population	32	44
		%	49.2	67.7
	Spain	Population	46	19
		%	76.7	31.7
	Poland	Population	42	44
		%	60.0	62.9
	Portugal	Population	41	21
		%	74.5	38.2
	Slovakia	Population	48	41
		%	68.6	58.6
	Sweden	Population	42	17
		%	84.0	34.0
	Hungary	Population	29	35
		%	52.7	63.6
	Italy	Population	34	40
		%	56.7	66.7
Total		Population	314	261
		%	64.7	53.8
Chi-square independence test			$\chi^2 = 27.36; p < 0.001$	$\chi^2 = 39.25; p < 0.001$

Table 8.11 Water savings by a country of residence

			Water savings	
			Yes	No
Country	The Czech Republic	Population	52	13
		%	80.0	20.0
	Spain	Population	55	5
		%	91.7	8.3
	Poland	Population	60	10
		%	85.7	14.3
	Portugal	Population	48	7
		%	87.3	12.7
	Slovakia	Population	53	17
		%	75.7	24.3
	Sweden	Population	44	6
		%	88.0	12.0
	Hungary	Population	47	8
		%	85.5	14.5
	Italy	Population	49	11
		%	81.7	18.3
Total		Population	408	77
		%	84.1	15.9
Chi-square independence test			$\chi^2 = 8.54; p = 0.288$	

In the vast majority of cases, people over the age of 35 (88%) saved water, compared to 78% of respondents in the age group up to 35 years. The difference is statistically significant ($p = 0.003$). Such a correlation was still found when respondents from Slovakia ($p = 0.002$) and Hungary ($p = 0.029$) were analyzed separately (Table 8.12).

Place of residence did not affect the fact of saving water in the case of all respondents ($p = 0.224$). Such an impact was found only in the case of the Czech respondents ($p = 0.013$), rural inhabitants (93% of them) were saving water more often than inhabitants of Czech cities (67% of them) (Table 8.13).

The respondents answered that they most often saved water by turning off the tap while brushing their teeth (70%) and checking and repairing leaking taps (62%), slightly less water was saved by starting the washing machine only when it was full (56%), taking a shower instead of taking a bath (53%), washing the dishes in the dishwasher (51%). The least frequently chosen form of saving was starting the dishwasher only when it was full (36%). The analysis by means of the Chi-square independence test did not show statistically significant differences between

Table 8.12 Water savings by a country of residence and age of respondents; % of people saving

		Age		Chi-square (χ^2; p)
		<35 yrs old	≥35 yrs old	
The Czech Republic	Population	19	32	0.34
	%	76.0%	82.1%	0.557
Spain	Population	18	37	0.11
	%	90.0%	92.5%	0.741
Poland	Population	27	33	0.77
	%	81.8%	89.2%	0.379
Portugal	Population	19	29	0.77
	%	82.6%	90.6%	0.379
Slovakia	Population	15	37	9.37
	%	55.6%	88.1%	0.002
Sweden	Population	17	27	0.06
	%	89.5%	87.1%	0.802
Hungary	Population	16	31	4.78
	%	72.7%	93.9%	0.029
Italy	Population	22	27	0.00
	%	81.5%	81.8%	0.973
Total	Population	153	253	8.85
	% of Age	78.1%	88.2%	0.003

countries in the frequency of indicating individual saving methods. The values of Chi-squared coefficients together with levels of significance and percentage distributions of responses are presented in Table 8.14.

An average European consumes 110 L of water a day on average. 70% of respondents knew that half of the water consumed can be replaced by lower quality water, i.e., rainwater or gray water. Most of them were from Spain (83%), Portugal (78%), Sweden (78%), Poland (71%), slightly less surveyed from Italy and the Czech Republic (62%), the smallest percentage of respondents with such knowledge came from Hungary (49%). The difference between the groups is statistically significant ($p = 0.002$) (Table 8.15).

Chi-square analysis showed a statistically significant difference in the distribution of responses between respondents from countries with the moderate climate, and respondents from countries with the Mediterranean climate ($p = 0.049$). People from Mediterranean countries (75%) knew more often that water used could be replaced with lower quality water than people from countries with moderate climate (67%) (Table 8.16).

The most important questions in the survey, which were to indicate the reasons for the low interest in alternative plumbing installations were formulated as follows:

Table 8.13 Water savings by a country and a place of residence;% of people saving

		Place of residence		Chi-square (χ^2; p)
		Country	City	
The Czech Republic	Population	28	24	6.19
	%	93.3%	68.6%	0.013
Spain	Population	20	35	0.54
	%	95.2%	89.7%	0.463
Poland	Population	23	37	0.01
	%	85.2%	86.0%	0.920
Portugal	Population	21	27	0.58
	%	91.3%	84.4%	0.447
Slovakia	Population	15	38	0.15
	%	78.9%	74.5%	0.700
Sweden	Population	15	29	0.00
	%	88.2%	87.9%	0.971
Hungary	Population	19	28	0.02
	%	86.4%	84.8%	0.876
Italy	Population	16	33	1.85
	%	72.7%	86.8%	0.173
Total	Population	157	251	1.48
	% of Age	86.7%	82.6%	0.224

1. Would you be afraid of using gray water in your household for the following purposes (toilet flushing, washing, garden watering, cleaning works, car washing)?
2. Would you like to use the gray water system in your household? If NO, please indicate for what reasons (hygiene reasons, high investments).
3. Would you be afraid of using rainwater in your household for the following purposes (toilet flushing, washing, garden watering, cleaning works, car washing)?
4. Would you like to use a rainwater harvesting system in your household? If NO, please indicate for what reasons (hygiene reasons, high investments).

A large percentage of respondents were afraid of using gray water in their household (60%). Respondents from Spain (58%), Italy (57%), and Portugal (56%) had the most concerns, respondents from Slovakia (81%), the Czech Republic (79%), and Hungary (73%) had the greatest concerns. The analysis using the Chi-square test showed that the difference was statistically significant ($p < 0.001$) (Table 8.17 and 8.18).

The most common concerns about the use of gray water had respondents from countries with significant water resources, over 10,000 m^3 per inhabitant (71% of them), less frequently respondents living in countries with less water resources (54%)

Table 8.14 Ways to save water, % of indications, divided into the country of residence

			Turning off the tap while brushing teeth	Washing dishes in the dishwasher	Starting the washing machine only when it is full	Starting the dishwasher only when it is full	Checking and repairing leaking taps	Taking a shower instead of taking a bath
Country	The Czech Republic	Population	44	37	36	30	34	28
		%	67.7	56.9	55.4	46.2	52.3	43.1
	Spain	Population	43	37	33	21	35	30
		%	71.7	61.7	55.0	35.0	58.3	50.0
	Poland	Population	47	36	41	30	39	37
		%	67.1	51.4	58.6	42.9	55.7	52.9
	Portugal	Population	40	29	35	17	40	34
		%	72.7	52.7	63.6	30.9	72.7	61.8
	Slovakia	Population	42	29	32	24	37	35
		%	60.0	41.4	45.7	34.3	52.9	50.0
	Sweden	Population	38	25	28	20	36	31
		%	76.0	50.0	56.0	40.0	72.0	62.0
	Hungary	Population	41	27	35	17	39	32
		%	74.5	49.1	63.6	30.9	70.9	58.2
	Italy	Population	44	27	33	15	40	27
		%	73.3	45.0	55.0	25.0	66.7	45.0
Total		Population	339	247	273	174	300	254
		%	69.9	50.9	56.3	35.9	61.9	52.4
Chi-square independence test			$\chi^2 = 5.75; p = 0.570$	$\chi^2 = 7.24; p = 0.404$	$\chi^2 = 5.85; p = 0.558$	$\chi^2 = 9.20; p = 0.239$	$\chi^2 = 13.78; p = 0.055$	$\chi^2 = 8.43; p = 0.296$

and respondents from countries with climate moderate (70%) than respondents from Mediterranean countries (43% of them). The analysis using the Chi-square test showed that the difference is statistically significant ($p < 0.001$) (Table 8.19).

The biggest concern was the use of gray water for washing (55%) and cleaning works (38%). Gray water was used to water the garden (24%), flush the toilet (20%), and wash the car (16%).

Statistical analysis showed a statistically significant difference between countries in the distribution of individual responses. The Hungarians (38%), the Czechs (25%), least often the Poles (11%), the Italians (12%), and the Portuguese (13%) were worried about using gray water, water to flush the toilet ($p = 0.005$). The Hungarians (73%) and the Slovaks (70%) were most afraid of washing in water from treated gray water; the least concerns were in Spain (42%), the Italians (43%), and the Portuguese (44%) ($p < 0.001$). The Italians, the Spaniards (15%) and the Portuguese (15%) would not be afraid to water the gray water, while the worst were the Slovaks (30%), the Swedes (30%), and the Hungarians (32%) ($p = 0.024$). The Italians (7%)

Table 8.15 A statistical European consumes 110 L of water per day. Does the respondent know that almost half of this value can be replaced with lower quality water: rainwater or gray water. Distribution of answers divided into the country of residence

			Response	
			Yes	No
Country	The Czech Republic	Population	40	25
		%	61.5	38.5
	Spain	Population	50	10
		%	83.3	16.7
	Poland	Population	50	20
		%	71.4	28.6
	Portugal	Population	43	12
		%	78.2	21.8
	Slovakia	Population	51	19
		%	72.9	27.1
	Sweden	Population	39	11
		%	78.0	22.0
	Hungary	Population	26	27
		%	49.1	50.9
	Italy	Population	39	21
		%	65.0	35.0
Total		Population	338	145
		%	70.0	30.0
Chi-square independence test			$\chi^2 = 22.70; p = 0.002$	

Table 8.16 A statistical European consumes 110 L of water per day. Does the respondent know that almost half of this value can be replaced with lower quality water: rainwater or gray water. Distribution of answers divided into the country climate

			Response	
			Yes	No
Country climate	Moderate	Population	206	102
		%	66.9	33.1
	Mediterranean	Population	132	43
		%	75.4	24.6
Chi-square independence test			$\chi^2 = 3.88; p = 0.049$	

Table 8.17 Would the respondent be afraid of using gray water in their household

			Response	
			Yes	No
Country	The Czech Republic	Population	51	14
		%	78.5	21.5
	Spain	Population	25	35
		%	41.7	58.3
	Poland	Population	41	29
		%	58.6	41.4
	Portugal	Population	24	31
		%	43.6	56.4
	Slovakia	Population	57	13
		%	81.4	18.6
	Sweden	Population	27	23
		%	54.0	46.0
	Hungary	Population	40	15
		%	72.7	27.3
	Italy	Population	26	34
		%	43.3	56.7
Total		Population	291	194
		%	60.0	40.0
Chi-square independence test			$\chi^2 = 48.63; p < 0.001$	

Table 8.18 Would the respondent be afraid of using gray water in their household, a division according to the country's water resources

			Response	
			Yes	No
Country water resources (calculated per one inhabitant)	<10 000 m^3	Population	167	143
		%	53.9	46.1
	>10 000 m^3	Population	124	51
		%	70.9	29.1
Chi-square independence test			$\chi^2 = 13.45; p < 0.001$	

and the Spaniards (10%) were not afraid of washing the car with pretreated gray water, while the Poles (24%) and the Slovaks (24%) were worried ($p = 0.016$). The Czechs (52%) and the Slovaks (49%) were most afraid to use gray wastewater for cleaning works, the Portuguese (20%), the Spaniards (25%), and the Italians (27%) had the least concerns ($p < 0.001$) (Table 8.20).

Table 8.19 Would the respondent be afraid of using gray water in their household, a division by country climate

			Response	
			Yes	No
Country climate	Moderate	Population	216	94
		%	69.7	30.3
	Mediterranean	Population	75	100
		%	42.9	57.1
Chi-square independence test			$\chi^2 = 33.53; p < 0.001$	

Table 8.20 Gray water was used, biggest concern,% of answer "yes"

			Toilet flushing	Washing	Garden watering	Car washing	Cleaning works
Country	The Czech Republic	Population	16	41	19	13	34
		%	24.6	63.1	29.2	20.0	52.3
	Spain	Population	11	25	9	6	15
		%	18.3	41.7	15.0	10.0	25.0
	Poland	Population	8	35	17	17	25
		%	11.4	50.0	24.3	24.3	35.7
	Portugal	Population	7	24	8	4	11
		%	12.7	43.6	14.5	7.3	20.0
	Slovakia	Population	16	49	21	17	34
		%	22.9	70.0	30.0	24.3	48.6
	Sweden	Population	11	27	15	9	19
		%	22.0	54.0	30.0	18.0	38.0
	Hungary	Population	21	40	18	7	29
		%	38.2	72.7	32.7	12.7	52.7
	Italy	Population	7	26	7	4	16
		%	11.7	43.3	11.7	6.7	26.7
Total		Population	97	267	114	77	183
		%	20.0	55.1	23.5	15.9	37.7
Chi-square independence test			$\chi^2 = 20.45; p = 0.005$	$\chi^2 = 26.27; p < 0.001$	$\chi^2 = 16.17; p = 0.024$	$\chi^2 = 17.23; p = 0.016$	$\chi^2 = 29.39; p < 0.001$

When analyzing the answers to the question about the possibility of using rainwater in households, it was noted that 42% of respondents had concerns. The biggest concerns were the Slovaks (63%), the Czechs (57%), and the Hungarians (44%), while the youngest were the Poles (33%), the Portuguese (33%), the Spaniards (33%), and

Table 8.21 Would the respondent be afraid of using rainwater in their household

			Response	
			Yes	No
Country	The Czech Republic	Population	37	28
		%	56.9	43.1
	Spain	Population	20	40
		%	33.3	66.7
	Poland	Population	23	47
		%	32.9	67.1
	Portugal	Population	18	37
		%	32.7	67.3
	Slovakia	Population	44	26
		%	62.9	37.1
	Sweden	Population	19	31
		%	38.0	62.0
	Hungary	Population	24	31
		%	43.6	56.4
	Italy	Population	20	40
		%	33.3	66.7
Total		Population	205	280
		%	42.3	57.7
Chi-square independence test			$\chi^2 = 26.81; p < 0.001$	

Table 8.22 Would the respondent be afraid of using rainwater in their household, a division by country water resources

			Response	
			Yes	No
Country water resources (calculated per one inhabitant)	<10 000 m^3	Population	118	192
		%	38.1	61.9
	>10 000 m^3	Population	87	88
		%	49.7	50.3
Chi-square independence test			$\chi^2 = 6.22; p = 0.013$	

the Italians (33%). The analysis by means of the Chi-square dependency test showed that the difference was statistically significant ($p < 0.001$) (Table 8.21).

Most often, respondents from countries with large water resources, above 10,000 m^3 per inhabitant (50% of them), less frequently respondents from countries with less water resources (38%) and respondents from countries with moderate climate had the fear of using rainwater. (47%) than respondents from Mediterranean

Table 8.23 Would the respondent be afraid of using rainwater in their household, a division by country climate

			Response	
			Yes	No
Country climate	Moderate	Population	147	163
		%	47.4	52.6
	Mediterranean	Population	58	117
		%	33.1	66.9
Chi-square independence test			$\chi^2 = 9.34; p = 0.002$	

Table 8.24 Rainwater use, biggest concerns,% answer "yes"

			Toilet flushing	Washing	Garden watering	Car washing	Cleaning works
Country	The Czech Republic	Population	13	31	8	9	18
		%	20.0	47.7	12.3	13.8	27.7
	Spain	Population	8	20	2	2	10
		%	13.3	33.3	3.3	3.3	16.7
	Poland	Population	8	23	1	6	10
		%	11.4	32.9	1.4	8.6	14.3
	Portugal	Population	6	18	1	1	9
		%	10.9	32.7	1.8	1.8	16.4
	Slovakia	Population	13	37	6	10	26
		%	18.6	52.9	8.6	14.3	37.1
	Sweden	Population	5	19	0	2	9
		%	10.0	38.0	0.0	4.0	18.0
	Hungary	Population	11	24	0	3	12
		%	20.0	43.6	0.0	5.5	21.8
	Italy	Population	3	20	0	2	10
		%	5.0	33.3	0.0	3.3	16.7
Total		Population	67	192	18	35	104
		%	13.8	39.6	3.7	7.2	21.4
Chi-square independence test			$\chi^2 = 10.45; p = 0.165$	$\chi^2 = 11.74; p = 0.109$	$\chi^2 = 26.02; p < 0.001$	$\chi^2 = 15.81; p = 0.027$	$\chi^2 = 16.70; p = 0.019$

countries (33% of them). The analysis using the Chi-square test showed that the difference was statistically significant ($p = 0.013; p = 0.002$, respectively) (Tables 8.22 and 8.23).

Table 8.25 Would the respondent like to use a system of using rainwater in their household, a division of responses by a country

			Response	
			Yes	No
Country	The Czech Republic	Population	45	20
		%	69.2	30.8
	Spain	Population	41	19
		%	68.3	31.7
	Poland	Population	58	12
		%	82.9	17.1
	Portugal	Population	36	19
		%	65.5	34.5
	Slovakia	Population	46	24
		%	65.7	34.3
	Sweden	Population	30	20
		%	60.0	40.0
	Hungary	Population	31	24
		%	56.4	43.6
	Italy	Population	36	24
		%	60.0	40.0
Total		Population	323	162
		%	66.6	33.4
Chi-square independence test			$\chi^2 = 13.40; p = 0.063$	

Table 8.26 Would the respondent like to use a system of using rainwater in their household, a division of responses by a country water resources

			Response	
			Yes	No
Country water resources (calculated per one inhabitant)	<10 000 m^3	Population	216	94
		%	69.7	30.3
	>10 000 m^3	Population	107	68
		%	61.1	38.9
Chi-square independence test			$\chi^2 = 3.66;$ $p = 0.056$	

The biggest concerns in the use of rainwater in the household were associated with its use for washing (40%) and cleaning work (21%), less often for toilet flushing (14%). Chi-square analysis showed a statistically significant difference between the respondents from individual countries. The Czechs (12%) had the greatest fears regarding garden watering, Swedes, Hungarians, and Italians did not have such fears

Table 8.27 Would the respondent like to use a system of using rainwater in their household, a division of responses by a country climate

			Response	
			Yes	No
Country climate	Moderate	Population	210	100
		%	67.7	32.3
	Mediterranean	Population	113	62
		%	64.6	35.4
Chi-square independence test			$\chi^2 = 0.51; p = 0.477$	

Table 8.28 Would the respondent like to use the rainwater harvesting system in his household by a country of residence and age of the respondents, % of respondents indicating "yes"

		Age		Chi-square ($\chi^2; p$)
		<35 yrs old	≥35 yrs old	
The Czech Republic	Population	22	22	7.08
	%	88.0%	56.4%	0.008
Spain	Population	12	29	0.96
	%	60.0%	72.5%	0.326
Poland	Population	29	29	1.11
	%	87.9%	78.4%	0.292
Portugal	Population	16	20	0.30
	%	69.6%	62.5%	0.587
Slovakia	Population	18	27	0.04
	%	66.7%	64.3%	0.839
Sweden	Population	12	18	0.13
	%	63.2%	58.1%	0.721
Hungary	Population	12	19	0.05
	%	54.5%	57.6%	0.824
Italy	Population	15	21	0.40
	%	55.6%	63.6%	0.525
Total	Population	136	185	1.27
	% of Age	69.4%	64.5%	0.260

at all ($p < 0.001$). Washing the car with rainwater was the most feared by the Czechs (14%) and the Slovaks (14%), while the Portuguese (2%), the Spaniards (3%), and the Italians (3%) were not afraid ($p = 0.027$). The Czechs (28%) and Slovaks (37%) were the most afraid to use rainwater for cleaning work, the Poles (14%), the Spaniards (17%), and the Italians (17%) were the least afraid ($p = 0.019$) (Table 8.24).

The next two questions asked about the willingness to use the rainwater harvesting system (RWH) and graywater recycling system (GWR) in the household. A rainwater

Table 8.29 Reasons why respondents would like to use the rainwater system in their household,% of responses

			Increased investments	Hygienic reasons	Already applied
Country	The Czech Republic	Population	12	8	0
		%	18.5	12.3	0.0
	Spain	Population	10	13	0
		%	16.7	21.7	0.0
	Poland	Population	9	3	1
		%	12.9	4.3	1.4
	Portugal	Population	12	9	0
		%	21.8	16.4	0.0
	Slovakia	Population	12	15	0
		%	17.1	21.4	0.0
	Sweden	Population	10	14	0
		%	20.0	28.0	0.0
	Hungary	Population	14	16	0
		%	25.5	29.1	0.0
	Italy	Population	14	15	0
		%	23.3	25.0	0.0
Total		Population	93	93	1
		%	19.2	19.2	0.2
Chi-square independence test			$\chi^2 = 4.59; p = 0.709$	$\chi^2 = 20.06; p = 0.005$	$\chi^2 = 5.94; p = 0.547$

harvesting system would apply 67% of all respondents. The Poles most often wanted to use it (83%). In the case of other countries, the percentage of respondents wishing to use the rainwater system was similar and ranged from 56 to 69%. The analysis using the Chi-square test showed no statistically significant differences ($p = 0.063$) (Table 8.25). In addition, statistical analysis using the Chi-square test showed that the distribution of responses was related to the country's water resources ($p = 0.056$) and its climate ($p = 0.477$) (Tables 8.26 and 8.27).

In the case of all respondents, it was not found that the respondent's age was associated with the desire to use (RWH) the system ($p = 0.260$). However, this correlation was found for those surveyed from the Czech Republic ($p = 0.008$). Rainwater harvesting system would be primarily used by people under the age of 35 (88%), less often older respondents (56%) (Table 8.28). The same percentage of respondents indicated that the reason why the respondents do not want to use the rainwater harvesting system was both increased investment (19%) and hygiene (19%). The Chi-square analysis showed differences in the frequency of indicating hygiene reasons between respondents from individual countries ($p = 0.005$). The

Table 8.30 Would the respondent like to use the graywater system in their household, a division of responses by a country

			Response	
			Yes	No
Country	The Czech Republic	Population	26	39
		%	40.0	60.0
	Spain	Population	33	27
		%	55.0	45.0
	Poland	Population	39	31
		%	55.7	44.3
	Portugal	Population	28	27
		%	50.9	49.1
	Slovakia	Population	29	41
		%	41.4	58.6
	Sweden	Population	19	31
		%	38.0	62.0
	Hungary	Population	15	40
		%	27.3	72.7
	Italy	Population	39	21
		%	65.0	35.0
Total		Population	228	257
		%	47.0	53.0
Chi-square independence test			$\chi^2 = 24.19; p = 0.001$	

Table 8.31 Would the respondent like to use the graywater system in their household, a division of responses by a country water resources

			Response	
			Yes	No
Country water resources (calculated per one inhabitant)	<10 000 m^3	Population	165	145
		%	53.2	46.8
	>10 000 m^3	Population	63	112
		%	36.0	64.0
Chi-square independence test			$\chi^2 = 13.32; p < 0.001$	

Table 8.32 Would the respondent like to use the graywater system in their household, a division of responses by a country climate

			Response	
			Yes	No
Country climate	Moderate	Population	128	182
		%	41.3	58.7
	Mediterranean	Population	100	75
		%	57.1	42.9
Chi-square independence test			$\chi^2 = 11.28; p = 0.001$	

Table 8.33 Reasons why respondents would like to use the graywater system in their household, % of responses

			Increased investments	Hygienic reasons
Country	The Czech Republic	Population	12	27
		%	18.5	41.5
	Spain	Population	18	16
		%	30.0	26.7
	Poland	Population	13	23
		%	18.6	32.9
	Portugal	Population	16	16
		%	29.1	29.1
	Slovakia	Population	15	34
		%	21.4	48.6
	Sweden	Population	14	26
		%	28.0	52.0
	Hungary	Population	14	33
		%	25.5	60.0
	Italy	Population	13	17
		%	21.7	28.3
Total		Population	115	192
		%	23.7	39.6
Chi-square independence test			$\chi^2 = 5.15; p = 0.642$	$\chi^2 = 26.50; p < 0.001$

most frequently mentioned were the Hungarians (29%), the Swedes (28%), the Italians (25%), the Spaniards (22%), the Slovaks (21%), the least frequently the Poles (4%) and the Czechs (12%) (Table 8.29).

Referring to the possibility of using the graywater recycling system (GWR) in the household, it was observed that the respondents were more willing to install it than

Table 8.34 Would cofinancing to install these systems encourage the respondent to use them in their household

			Response	
			Yes	No
Country	The Czech Republic	Population	43	22
		%	66.2	33.8
	Spain	Population	44	16
		%	73.3	26.7
	Poland	Population	59	11
		%	84.3	15.7
	Portugal	Population	41	14
		%	74.5	25.5
	Slovakia	Population	53	17
		%	75.7	24.3
	Sweden	Population	41	9
		%	82.0	18.0
	Hungary	Population	37	18
		%	67.3	32.7
	Italy	Population	46	14
		%	76.7	23.3
Total		Population	364	121
		%	75.1	24.9
Chi-square independence test			$\chi^2 = 9.21; p = 0.238$	

was the case with the rainwater harvesting system. 47% of respondents wanted to use the graywater system in their households. The most frequent respondents from Italy (65%), Poland (56%), and Spain (55%), the least willing to use the system were among the respondents from Hungary (27%) and Sweden (38%). The difference between the groups is statistically significant ($p = 0.001$) (Table 8.30).

Most often, GWR would like to be used by respondents from countries with low water resources (53%), less frequently respondents from countries with larger water resources (36%) and respondents from countries with Mediterranean climate (57%) than those from countries with moderate climate (41%). The difference between the groups is statistically significant ($p < 0.001$) (Tables 8.31 and 8.32).

The main reason why the respondents would like to use the graywater system in their household is hygiene (40%), less often this reason was increased investment expenditure (24%). The Chi-square analysis showed a difference between respondents from individual countries in the frequency of indicating hygiene reasons ($p < 0.001$). Most often they were mentioned by the Hungarians (60%), the Swedes (52%), the Spaniards (22%), the Slovaks (49%), least often the Spaniards (27%), and the Italians (28%) (Table 8.33).

Table 8.35 Would cofinancing to install these systems encourage the respondent to use them in his household by country of residence and age of the respondents,% of respondents indicating "yes"

		Age		Chi-square (χ^2; p)
		<35 yrs old	≥35 yrs old	
The Czech Republic	Population	17	25	0.10
	%	68.0%	64.1%	0.749
Spain	Population	14	30	0.17
	%	70.0%	75.0%	0.680
Poland	Population	29	30	0.61
	%	87.9%	81.1%	0.435
Portugal	Population	17	24	0.01
	%	73.9%	75.0%	0.927
Slovakia	Population	20	32	0.04
	%	74.1%	76.2%	0.842
Sweden	Population	15	26	0.19
	%	78.9%	83.9%	0.660
Hungary	Population	16	21	0.50
	%	72.7%	63.6%	0.481
Italy	Population	24	22	4.10
	%	88.9%	66.7%	0.043
Total	Population	152	210	1.19
	% of Age	77.6%	73.2%	0.275

Taking into account the activities of many supporting countries, alternative sources of water were used, the respondents were also asked whether the cofinancing would be an incentive for them to use a system of using rainwater and recycling gray water in their households. The vast majority of all respondents (75%) answered "yes". The analysis using the Chi-square test showed no statistically significant differences in the distribution of responses between respondents from different countries ($p = 0.238$) (Table 8.34).

It was not also found that the division of responses was related to the age of the respondent when all respondents were considered ($p = 0.275$). Such a difference was found in the case of Italian respondents, where younger respondents would be more often encouraged by funding (89% of them) than older respondents (67%) ($p = 0.043$) (Table 8.35).

The research results obtained showed that more than half of the respondents thought that there was a shortage of potable water in their country. This was especially noticeable in countries with large potable water supplies, such as Spain and Portugal, where over 76% of respondents indicated a problem with the availability of potable water. When analyzing the results, the influence of a given country's climate on the inhabitants' responses was also noticed. Water selection was noticed most often by people from the countries with Mediterranean climate (67%) than people from countries with moderate climate (41%). In a situation where all respondents were considered, it was not found, however, that the place of residence (city or village) significantly differentiated the perception of water supply. Similar observations are presented in the publication by Garcia-Cuerva et al. (2016).

It is worth noting that the majority of respondents (84%) save water in their households through daily activities, such as turning off the tap while brushing teeth (70%), starting the washing machine only when it is full (56%), taking a shower instead of taking a bath (53%), and washing dishes in the dishwasher (51%). The analysis did not show statistically significant differences between countries in the frequency of indicating individual ways of saving.

Most respondents would be afraid of using gray water (60%), while they were more likely to use rainwater in their households, which was indicated by 58% of respondents. In both cases, the Slovaks, the Czechs, and the Hungarians had the greatest fears. This may indicate that in countries like Spain and Portugal, residents are more aware and friendly about alternative water sources. This is certainly influenced by the information campaign and the promotional law changed in recent years, and in some cases mandatory use of conventional water systems (Domènech and Saurí 2010).

On the basis of the results obtained, it was also observed that the climate and the water resources of a given country had an impact on the responses in the use of gray water and rainwater in households. The most fears aroused the use of these water sources among the inhabitants of countries with low water resources, especially those located in the region where the Mediterranean climate occurs. It was also noted that in the case of both sources, the biggest concerns, caused by hygiene reasons, aroused them were used for washing and cleaning works. Most willingly, respondents would be willing to replace potable water with pretreated gray water or rainwater used to flush toilets and water the garden.

In addition to hygiene reasons, as a reluctance to implement these systems, the respondents indicated increased investments. Over 75% of respondents said that cofinancing for investments would be a great incentive for them to use the systems examined. Societies' beliefs about the unprofitability of alternative water systems and their hygiene concerns are mainly caused by the lack of social campaigns promoting these systems and informing about financial and environmental benefits resulting from their use. A change in the law could also contribute to the increase in interest in these systems, funding by governments and nongovernmental institutions to install these systems was offered, and lower taxes were introduced for people who would apply such solutions in their buildings.

References

Arrkelin D (2014) Using SPSS to understand research and data analysis. Psychology curricular materials. Book 1. Valparaiso University

Dolnicar S, Schäfer A (2009) Desalinated versus recycled water: public perceptions and profiles of the accepters. J Environ Manag 90(2):888–900

Dolnicar S, Hurlimann A, Nghiem L (2010) The effect of information on public acceptance—the case of water from alternative sources. J Environ Manag 91(6):1288–1293

Domènech L, Saurí D (2010) Socio-technical transitions in water scarcity contexts: public acceptance of greywater reuse technologies in the Metropolitan Area of Barcelona. Resour Conserv Recycl 55:53–62

Fielding KS, Dolnicar S, Schultz T (2018) Public acceptance of recycled water. Int J Water Resour Dev 1:1–36

Garcia-Cuerva L, Berglund E, Binder A (2016) Public perceptions of water shortages, conservation behaviors, and support for water reuse in the U.S. Resour Conserv Recycl 113:106–115

Gu Q, Chen Y, Pody R, Cheng R, Zheng X, Zhang Z (2015) Public perception and acceptability toward reclaimed water in Tianjin. Resour Conserv Recycl 104:291–299

Hurlimann A, Dolnicar S (2010) When public opposition defeats alternative water projects—the Case of Toowoomba Australia. Water Res 44(1):287–297

Ong M, Puteh F (2017) Quantitative data analysis: choosing between SPSS, PLS and AMOS in social science research. Int Interdiscip J Sci Res 3:14–25

Skoien G, Gardner T (2008) Decentralised water supplies. South East Queensland householders' experience and attitudes. Water 35:17–22

Chapter 9
Summary and Final Conclusions

Water is one of the most important resources of the natural environment that underpins human existence. However, over the years, freshwater resources have been overexploited as a result of anthropogenic activities. This has led, in many regions of the world, to a state where their quantity and quality is not adequate to ensure proper social and economic development. Occurring water shortages, which are caused not only by improper management of its resources but also by the growing demand for it, a changing climate and an intensive process of urbanization are now becoming one of the biggest global problems. Therefore, actions are taken to introduce and implement sustainable water and sewage management. It is a strategy whose primary goal is to maintain water resources in good condition and their exploitation at such a level that economic and social development for current and future generations is possible.

Modern water management should be based on its sustainable consumption, based not only on available freshwater resources but also on alternative water sources, which are rainwater and gray water. Therefore, it is necessary to conduct research, analysis, and the impact assessments of the use of unconventional water systems in buildings in the financial, social, and environmental context.

Considering the above, scientific research was undertaken to determine the efficiency of rainwater harvesting systems (RWHS) and graywater recycling systems (RWRS) in single-family buildings located in selected European cities. The impact of many technical parameters of these systems and climatic conditions on the effectiveness of tap water saving and the resulting financial benefits were investigated. Research has also been conducted to explore the social aspect, which is often decisive in the implementation of alternative solutions.

On the basis of the research and analyses carried out in the field of research, the following **general conclusions** can be drawn:

- carrying out a detailed technical and economic analysis taking into account the technical parameters, local climatic conditions, and financial parameters, allows one to choose the optimal variant of the water installation in the building that allows achieving the best financial indicators in the long term,

© Springer Nature Switzerland AG 2020
A. Stec, *Sustainable Water Management in Buildings*,
Water Science and Technology Library 90,
https://doi.org/10.1007/978-3-030-35959-1_9

- the use of Life Cycle Cost analysis in the decision-making process regarding the selection of the most favorable investment variant ensures that the financially correct choice is made. On the other hand, adopting a solution based only on initial investment outlays may lead to making incorrect decisions from the investor's point of view, because in many cases, investment outlays constitute a small part of the costs incurred throughout the lifetime of the facility,
- decentralized systems that are considered rainwater harvesting systems and gray-water recycling systems allow not only to save water, but also allow you to partially or completely become independent of centralized water suppliers, which in turn can increase water safety,
- the implementation of RWHS and GWRS can contribute significantly to sustainability at the building level,
- gray water, despite the fact that in most cases their use is not financially viable, can be a valuable source of water, which, unlike rainwater which is dependent on climate conditions, is available in the building in constant quantities throughout the year. This may be important especially in regions with high rainfall variability throughout the year, long periods of no-rain or winter months with negative temperatures,
- the roof surface, the demand for non-potable water, the amount of rainfall, and the capacity of the rainwater storage tank have a decisive impact on the efficiency of rainwater harvesting system,
- determination of volumetric reliability for various tank capacities allows to determine the optimal tank size for variable demand for water of lower quality in the building, which allows achieving the greatest tap water savings,
- the impact of a dry and wet year on volumetric reliability is particularly evident in locations with a hot climate and mild winters, where rainfall flows from the roof to the reservoir throughout the year or most of the year. In the case of areas with a moderate climate, where over many years there are no large differences in the sum of annual rainfall, and RWHS in winter, due to negative temperatures, does not work, this impact is insignificant,
- the financial efficiency of the examined systems located in different parts of Europe, in addition to technical parameters and climatic conditions, was significantly influenced by the purchase price of water from the water supply network and for the discharge of wastewater into the sewage system, as well as the annual increase in these prices,
- residents of countries with small water resources are more aware and friendly about the use of alternative water sources, which could have been influenced by the law changed in these countries in recent years to promote, and sometimes mandate the use of unconventional water systems and social information campaigns,
- in most of the cases studied, rainwater as an alternative source of water is acceptable to the public, but still, for hygiene reasons, its use for washing is the biggest concern. Preferably, it would be used to flush toilets and water the garden.

The research results obtained in the scope of the possibility of using rainwater and gray water in single-family buildings located in selected European cities also allowed the formulation of a number of **specific conclusions of cognitive and application significance**.

- the highest volumetric reliability Vr in the case of using rainwater only for toilet flushing (Variant 1) was characterized by the rainwater harvesting system (RWHS) located in Rome (99%) and Lisbon (98%), where rainwater flows due to warm winters to the tank practically all year round and only a small part of it is discharged into the sewage system. Such results were also influenced by the amount of rainfall whose average over the period of the analyzed 10 years was the highest and amounted to 693 mm and 687 mm, respectively. In the situation of the RWHS location in Madrid, despite the fact that the rainfall was the lowest for 10 years, the efficiency of this system was at the level of 91%. Such a high Vr value was mainly due to the rainwater flowing into the rainwater harvesting system for about 10 months during the year. Lower rainwater harvesting system efficiency of up to 84% was obtained for the RWHS location in Budapest. In contrast to the cases described above, the differences between the level of volumetric reliability for different water needs and different roof sizes were small or nonexistent. Maximum Vr was achieved for much smaller tank volumes than in Rome, Lisbon, and Madrid. It was 3 m^3, 5 m^3, and 9 m^3, for two persons, three persons, and four persons, respectively. Volumetric reliability Vr of rainwater harvesting systems located in Warsaw, Stockholm, and Bratislava reached a maximum of 80%. No significant differences between Vr were observed for these locations for different needs of rainwater resulting from the variable number of inhabitants.
- the highest efficiency of RWHS for Variant 2 assuming the use of rainwater for toilet flushing and laundry was obtained, as well as for Variant 1, in the case of its location in Rome and Lisbon. Volumetric reliability values for both locations were at a similar level and differed by several percents. Comparing these results with the Vr values obtained for Variant 1, it was found that the largest decrease in volumetric reliability is observed when rainwater flows from a 100 m^2 roof. This is due to the smaller amount of rainwater flowing into the tank and the increased demand for rainwater in Variant 2. The larger the roof, the lower the Vr in Variant 2 compared to Variant 1 is smaller. It was also noted that in order to achieve maximum efficiency of RWHS, it is necessary to use larger tank capacities than in the situation where rainwater is used only for flushing toilets. Similar trends in the results in Variant 2 were also obtained for the rainwater harvesting system located in Madrid. The maximum system efficiency for two people was 77%, 88%, and 89%, for a 100 m^2, 150 m^2, and 200 m^2 roof, respectively. For a larger number of people and thus a greater demand for water for flushing toilets and laundry, volumetric reliability was lower, in some cases by up to 18%. Comparing the Vr values obtained for this location in Variant 1 and Variant 2, the smallest differences are observed, as in the case of Rome and Lisbon, for larger roof areas. When analyzing the results for the location of rainwater harvesting systems in Warsaw, Bratislava, and Stockholm, it was noted that the maximum volumetric reliability values obtained in Variant 2 are very similar to Vr in Variant 1 and are 80%. This is especially visible in cases where two people use the installation.

- unlike Variant 1 and Variant 2 in the third variant, in which rainwater was used to flush toilets, wash, and water the garden, rainwater harvesting systems located in Rome, Lisbon, and Madrid were characterized by the lowest efficiency among all analyzed locations. The lowest volumetric reliability was obtained for Madrid, where the lowest rainfall occurred during the year. For two system users, the maximum Vr was 29%, 37%, and 42%, respectively, for a roof area of 100 m^2, 150 m^2 and 200 m^2. For three and four people, the decrease in volumetric reliability compared to the Vr level for two people was insignificant, in the range of 2–4%. The location of RWHS in Rome and Lisbon would allow it to achieve its efficiency of around 50%. The impact of the number of people on the final Vr was small. Very similar test results were obtained for other RWHS locations in Warsaw, Bratislava, Stockholm, Prague, and Budapest. The maximum volumetric reliability for these cities was around 90% for the largest roof area. In the case of this installation variant, the impact of the roof surface on the efficiency of RWHS was noticeable and, for example, for a 100 m^2 roof and two system users, the Vr decrease was even 20%. The number of inhabitants and the associated demand for rainwater affected volumetric reliability more than the location of the system in Rome, Lisbon, and Madrid. This was due to the fact that the amount of water required to water the garden was smaller than in the case of these three locations and constituted a smaller share in the total daily water demand in Variant 3.
- the highest Life Cycle Cost (LCC) value several times exceeding the value of these costs for other locations was obtained when the alternative installation systems considered were located in Lisbon. However, the results obtained showed that for none of the analyzed computational cases, the traditional installation solution (Variant 0) is not the optimal solution in financial terms. This is due to the fact that the operating costs associated with the operation of Variant 0 over a 30-year period are greater than for the use of individual installation systems powered from alternative water sources, despite the fact that they require higher investment outlays. Regardless of the number of inhabitants, the most advantageous in financial terms would be to use a variant in which both gray water and rainwater (Variant 5) are used, and the optimal tank capacity in RWHS was 11 m^3. For this capacity of the tank, this variant was characterized by LCC costs lower by almost 43,000 euros compared to Variant 0, despite the fact that the investment expenditure for the implementation of Variant 5 was five times higher than in the case of a traditional installation solution.
- the highest value of the LCC indicator for Variant 0 was also obtained in the case of locations of investment variants under consideration in Madrid, but in a situation where the building was inhabited by three and four people. If the installation was used by two users, the least cost-effective variant was Variant 4, in which only the gray water was the alternative source of water. It was observed that, similar to Lisbon for three and four inhabitants, the installation solution in which the rainwater harvesting system and the graywater recycling system were implemented was the most financially advantageous. Such a hybrid system allowed achieving the largest water savings in the analyzed period of 30 years, which in turn resulted in the lowest operating costs. These benefits were obtained for RWHS

equipped with a 9 m^3 tank. The situation looks different with the lowest number of inhabitants. If the installation is used by two people, the most profitable would be to use Variant 3 with a 5 m^3 tank, in which rainwater is used to flush toilets, wash, and water the garden.

- in the case of Prague, it would also be financially advantageous to use alternative water sources in a single-family house. However, for this location, in contrast to Lisbon and Madrid, the most profitable, regardless of the number of inhabitants, was variant 3 consisting of implementing RWHS for toilet flushing, washing, and watering the garden. If the system was used by two people, the Variant 3 with a 5 m^3 tank is financially optimal, while for a larger number of inhabitants, it is a RWHS equipped with a 7 m^3 tank.
- Variant 3 was also the solution with the lowest LCC costs for the building location in Rome. It was the only city among the eight analyzed that has annual fees for each cubic meter of rainwater discharged into the sewage system. As the results of the research have shown, it is the operating costs associated with the discharge of these waters to the drainage network that have decided about the profitability of using the variant with the rainwater harvesting system. The solution for the installation with the highest LCC costs, regardless of the number of people and the roof area, was variant 4 using gray water.
- the implementation of alternative water systems is not profitable for their location in Budapest, Bratislava, Warsaw, and Stockholm. The variant with the lowest costs for these locations was the traditional solution of the installation with water supply from the water supply network and sewage discharge to the sewage system (Variant 0). The results and hierarchy of profitability of individual variants were not significantly affected by either the number of users of the installation or the size of the roof area. The largest differences in the LCC ratio were observed when comparing Variant 0 and variants using graywater recycling systems, i.e., Variant 4 and Variant 5. The use of alternative water sources in single-family buildings in these cities was completely unprofitable, as water savings achieved during 30 years did not cover capital expenditure and operating costs caused by replacing filters and pumps in both systems. Comparing both alternative water sources, it can be stated that much better financial results were obtained for rainwater harvesting systems.
- extending the analysis period to 50 years did not increase the profitability of implementing RWHS and GWRS in those cities in which it was unprofitable for T = 30 years. As for T = 30 years, the largest LCC costs were obtained for variants with a wastewater recycling system (Variant 4 and Variant 5). Among the variants analyzed, the variant using rainwater for toilet flushing, washing, and watering the garden (Variant 3) was the most favorable compared to Variant 0. This is especially noticeable for four users of the installation. Such unfavorable results for variants with alternative water sources in buildings located in Budapest, Bratislava, Stockholm, and Warsaw, despite the extension of the LCC analysis period, are affected by relatively low purchase prices of water from the water supply network in these cities.

- shortening the T analysis time to 20 years had the greatest impact on the results obtained for Lisbon. This was due to the highest annual operating costs. For a longer analysis period, all alternative installation solutions were characterized by a lower level of LCC costs than the traditional installation variant, while for T = 20 years, the differences in the costs of individual variants were very small. An unfavorable change is especially noticeable in the case when the building was inhabited by four people, where the difference in the LCC value between Variant 0 and the most financially profitable variant has decreased from EUR 43,000 (T = 30 years) to about EUR 4,000 (T = 20 years). In addition, for the smallest number of system users, there was a change in the profitability hierarchy of the analyzed solutions and the variants with the graywater recycling system ceased to be financially advantageous variants.
- sensitivity analysis showed that the results obtained in the first stage can be considered correct, and the investment consisting of the implementation of RWHS and GWRS is slightly susceptible to changing individual parameters of the financial model. No significant changes in the profitability hierarchy of the analyzed installation variants were observed in most computational cases.
- hydrodynamic modeling of the selected urban catchment showed, depending on the number of RWHS used, the peak runoff was reduced in the range of 21% to even 100% and from 25% also to 100%, for variant 1 (implementation of RWHS in 50% of houses) and variant 2 (implementation of RWHS in 100% of houses) compared to variant 0 (drainage basin in the existing state without RWHS). A total reduction in the peak runoff was observed for negligible precipitation with an intensity of not more than 2 mm/hr. Precipitation with an intensity of 2 to 10 mm/hr caused an increase in the amount of wastewater flowing from the catchment and a reduction of this outflow by about 28%. On the other hand, with rainfall of more than 10 mm/hr, rainfall outflow was reduced by an average of about 26%. The application of RWHS in the catchment studied also resulted in an increase in the amount of rainwater that has infiltrated the soil through the green surfaces around the buildings. If in the catchment area, RWHS is implemented in 50% households (Variant 1), the infiltration will intensify from 9% even in some cases up to 400% compared to Variant 0. The average increase for this variant was about 30%, while for the variant 2, average 35%. However, the use of rainwater harvesting systems was not effective in reducing the occurrence of pressure flows in the analyzed sewage system. Many factors could influence such results including too few RWHS in relation to the total catchment area.
- surveys have shown that over 50% of respondents believe that there is a shortage of potable water in their country. 60% of respondents are afraid of using gray water in the household, and the Slovaks (81%), the Czechs (79%), and the Hungarians (73%) had the biggest concerns. For hygiene reasons, the use of this wastewater for washing (55%) and cleaning work (38%) was the biggest concern. When analyzing the answers to the question about the possibility of using rainwater in households, it was noticed that 42% of respondents expressed reluctance in this respect. As in the case of graywater recycling, respondents from Slovakia (63%), the Czech Republic (57%), and Hungary (44%) had the highest concerns, while the smallest

were the Portuguese (33%), the Spanish (33%), and the Italians (33%). The biggest concerns in the use of rainwater in the household were associated with its use for washing (40%) and cleaning work (21%), less often for toilet flushing (14%).

- the main reason why the respondents would not want to use unconventional water systems in their households was hygiene, as indicated by 40% of respondents when using gray water and 19% of those surveyed when using rainwater. In addition to hygiene reasons, as a reluctance to implement these systems, respondents indicated increased investments. However, the vast majority of all respondents (75%) replied that funding would be an incentive for them to use the graywater recycling system and rainwater harvesting system.

The results of the research also have a practical aspect and can be a guide for potential investors and operators to apply this type of system already at the investment planning stage. An additional impulse for their use, especially in locations where their use is not profitable, could be cofinancing from the state budget or from pro-ecological funds of international organizations, as is the case in many other countries. The conviction of societies about the unprofitability of alternative water systems and their hygiene concerns are primarily caused by the lack of social campaigns promoting these systems and informing about the financial and environmental benefits of their use. A change in the law and the introduction of appropriate education on water saving options could also contribute to the increase in interest in these systems.

Despite extensive research, taking into account technical, financial, and social aspects, no quantitative assessment of the impact of the analyzed systems on the natural environment and water resources has been made. Therefore, in further research, it is intended to perform the Life Cycle Assessment for various climatic regions, which is a cradle-to-grave analysis technique. Its results will allow one to design better alternative water systems and allow decision-makers to make more informed and sustainable decisions. In addition, a further direction of research will be to determine the impact of climate change in terms of rainfall on the efficiency of rainwater harvesting systems located in different climate conditions. A comparative analysis of decentralized water systems and the central water system in financial and environmental aspects is also intended.

CPSIA information can be obtained
at www.ICGtesting.com
Printed in the USA
LVHW080423131220
674041LV00003B/16

9 783030 359614